Photographs of
Interpreters

Photographs of Interpreters:

In Search of Lost Lives

By

John Milton

Cambridge
Scholars
Publishing

Photographs of Interpreters: In Search of Lost Lives

By John Milton

This book first published 2024

Cambridge Scholars Publishing

Lady Stephenson Library, Newcastle upon Tyne, NE6 2PA, UK

British Library Cataloguing in Publication Data
A catalogue record for this book is available from the British Library

ISBN (10): 1-0364-0654-7
ISBN (13): 978-1-0364-0654-7

TABLE OF CONTENTS

COPYRIGHT AND PERMISSIONS

The following photographs have been reprinted with permission as chapter images, and the respective copyright holders are explicitly stated below.

Chapter 1: (1) Archives diplomatiques, collection Collin de Plancy; (1a) Book cover reproduced with the permission of the Éditions Loubatières; (2) Public Domain; (3) Photo reproduced with the permission of the Bibliothèque nationale de France; (4) Photo reproduced with the permission of the Bibliothèque nationale de France; (5) ullsteinbild/TopFoto; (6 and 6a) Photo reproduced with the permission of the Bibliothèque nationale de France; (Ill. 1) by George Hodan CC0 Public Domain; (7) © Archives municipales de Saumur.

Chapter 2: (1,1a, 2 and 3) Nebraska State Historical Society Photograph Collections; (4, 5a and 5b) Courtesy of Hindman Auctions; (6) From the Bostwick-Frohardt Collection, Owned by KM3TV and on permanent loan to The Durham Museum; (Fig. 1) Photo by Dave Pape. Public Domain. (Fig. 2) Collection of Rob Niederman. Image used with permission. (7) "Lakota delegation to Washington May 13 1875".
https://quod.lib.umich.edu/p/pohrt/x-103/ste026_001. William L. Clements Library, University of Michigan Library Digital Collections. December 16, 2023. (8) Courtesy Western History Collections, Special Research Collections, University of Oklahoma Libraries, Walter Stanley Campbell Collection, Box: Photo C-84, item: 58.

Chapter 3: (1) [BRMI SPIRel888 1938] - Acervo do Museu do Índio/FUNAI – Brasil; (2) Acervo do Museu de Arqueologia e Etnologia da UFPR; (3) [BRMI SPIRel888 1287] - Acervo do Museu do Índio/FUNAI – Brasil; (Fig. 1 and 1a) [76.2.79d] - Acervo do Museu do Índio/FUNAI – Brasil; (4) [CRNV0051] - Acervo do Museu do Índio/FUNAI – Brasil; (5) [RR203] - Acervo do Museu do Índio/FUNAI – Brasil; (6) [SPIa1992] - Acervo do Museu do Índio/FUNAI – Brasil; (7, 8 and 9) [CRNV0242], [BRMI CRIcA2 723] and [BRMI CRIcA2 724] - Acervo do Museu do Índio/FUNAI – Brasil.

Chapter 4: (1) Photo by Ricardo Beliel, reproduced with permission; (2) Photo by Didier Descouens (CC-BY-SA-4.0). Public domain; (3) Photo by

Alonso de Mendoza. Public domain; (4) Photo by Ricardo Beliel, reproduced with permission; (5) Photograph by interpreters Vivian Celia Arango and Heriberto Rayón, reproduced with permission; (6) John van Hasselt – Corbis/Getty Images.

Chapter 5: (Fig. 1) George Henry Mason. Illustrated by J. Dadley. The punishments of China: illustrated by twenty-two engravings: with explanations in English and French. London: 1801. Plate VIII. PUNISHING AN INTERPRETER. Page 45. (1 and 2) John Thomson, Illustrations of China and its people. A series of two hundred photographs, with letterpress descriptive of the places and people represented, London, S. Low, Marston, Low, and Searle, 1873-74. Plate XVII, no. 34-35, volume 3. Page 106. (3) [CRNV0468] - Acervo do Museu do Índio/FUNAI – Brasil; (4 and 5) [BRMI CRIcA1 352] - Acervo do Museu do Índio/FUNAI – Brasil.

Chapter 6: (1) © Programa de Cooperación Hispano Peruano, Centro Amazónico de Antropología y Aplicación Práctica, Grupo Internacional de Trabajo sobre Asuntos Indígenas and Agencia Española de Cooperación Internacional para el Desarrollo; (2) Museum of Archaeology and Anthropology, Cambridge (P.100400.WHI); (3a) Museum of Archaeology and Anthropology, Cambridge (N.26773.WHI); (3b) © Programa de Cooperación Hispano Peruano, Centro Amazónico de Antropología y Aplicación Práctica, Grupo Internacional de Trabajo sobre Asuntos Indígenas and Agencia Española de Cooperación Internacional para el Desarrollo; (4) Museum of Archaeology and Anthropology, Cambridge (N.26769.WHI); (5) © Centro Amazónico de Antropología y Aplicación Práctica, Grupo Internacional de Trabajo sobre Asuntos Indígenas and Universidad Científica del Perú; (6) Photograph by Francois Dolmetsch, reproduced with permission; (7) Gómez, A. Lesmes, A. C. & Rocha, C. (1995). *Caucherías y conflicto colombo-peruano. Testimonios 1904-1934.* Bogotá: Disloque Editores. Coama.

Chapter 7: (1 and 2) Paul Schutzer/The LIFE Picture Collection/Shutterstock; (3) Gonzales/CPDOC JB/Folhapress.

Chapter 8: (1) The Washington Post/Getty Images; (2) Public domain. (3) Ben Martin/Archive Photos/Getty Images; (4) ZUMAPRESS.com/ Keystone Pictures USA/Age Photostock; (5) Photo from the Sukhodrev family archives.

Every effort has been made to trace copyright holders and to obtain their permission for the use of copyright material. We apologize for any errors or omissions and would be grateful to be notified of any corrections that should be incorporated in future editions of this book.

ACKNOWLEDGEMENTS

I would like to thank:

Jamilly Brandão for all her work on contacting the folders of rights to the photographs;

Renata Schinke for discovering and telling me about the photographs of the Young Well-dressed Indigenous Interpreter;

Telma São Bento Ferreira and Vera Lúcia Ramos at Editora Lexikos for publishing the Portuguese version of the book;

Marina Darmaros for helping me with the Russian interviews with Viktor Sukhodrev;

The USP postgraduate students who replied to the questionnaire on Chang and the Young Well-dressed Indigenous Interpreter;

Solange Pinheiro for her careful revision of the text;

Melina Valente for her eye-catching design of the cover;

Adam Rummens, Sophie Edminson, and Amanda Millar at Cambridge Scholars for all their hard work in the preparation of the book;

The Museu do Índio, Rio de Janeiro, and especially Luiza Zelesco, for allowing me to use a number of photos from the museum;

Ricardo Beliel for allowing me to use the very beautiful photo of the Korubo;

José Ribamar Bessa Freire, Pedro Libanio and Ana, curator of the "Civilized Indians" exhibition in Rio de Janeiro for the information on the Young Well-dressed Indigenous Interpreter;

Alisha Baginski at the Nebraska State Historical Society;

Céline Boudias of Bibliothèque nationale de France;

Didier Descouens for allowing me to reproduce the image he shared on Wikipedia;

Eric Duncan from Hindman Auctions;

François Dolmetsch for allowing me to use the photo of the *Colombia's Forgotten Frontier* cover.;

Haiden Nelson from The Durham Museum,

Helen Sutton at the Museum of Archaeology and Anthropology, Cambridge, UK;

Heriberto Rayón and Vivian Celia Arango for allowing me to use their APIC photo;

Jackie Reese from Western History Collections,

Judith Grima at Archives Municipales de Saumur

Marisa Basso from Folha Press;

Maryline Garrabos at Editions Loubatières;

Regiane Pelaquini from Museu de Arqueologia e Etnologia, Federal University of Paraná, Brazil; and

Rob Niederman for allowing me to use the photo of the 1860s Scovill camera from his personal archive.

Introduction

1. *Studium* and *Punctum*

Roland Barthes (1915-1980) uses the Latin word *studium* for a certain type of photograph: not exactly a "study" "but application to a thing, taste for someone, a kind of general, enthusiastic commitment, of course, but without special acuity" (Barthes 1981:26). Thus it is a somewhat superficial photograph, lame, "without salt and sugar", which delights and does not hurt, like a school book photograph, which presents the status quo, the official history of "good historical scene" through which the viewer will be able to "participate in the figures, the faces, the gestures, the settings, the actions" (Barthes 1981:26), seeing the scheme of the "Operator" of society; that is, "it is a kind of education" (Barthes 2015:28) in which the spectator will read and fraternize with the myths of the photographer though the spectator may not believe these myths, and, in the case of a conservative photo, they both will aim "at reconciling the Photograph with society", with the functions "to inform, to represent, to surprise, to cause, to signify, to provoke desire" (Barthes 1981:28). In Chapters 7 and 8, we will see how this *studium* relates to *LIFE* magazine.

The second element presented by Barthes breaks the *studium*. This is the *punctum,* like an arrow, a sting, a puncture, a small hole, a small stain, a small cut "which rises from the scene, shoots out of it like an arrow, and pierces me" and even punctuates (but also mortifies me, wounds me)" (Barthes 1981:26). The *studium* will frequently be "traversed, lashed, striped by a detail (*punctum*) which attracts or distresses me" (Barthes 1981:60), but when this fails to happen, we can call the photograph a "unary photography", that is, "no duality, no indirection, no disturbance" (Barthes 1981:61). Thus, without this *punctum*, the "unary photograph", the *studium*, would be a banal photograph, with a unit of composition, with just one unit, as in many reportage photographs, with no shock, injury or disturbance.

This division of Roland Barthes will be the basis for the analysis of a number of the photos in the following chapters.

2. Essence and Magic

One quality often present in commentaries on photographs and in interviews with photographers is that photography has a special ability to get to the "essence" of the scene or person being photographed. Siegfried Kracauer comments that nineteenth-century photographers such as Nadar (1820-1910), David Octavius Hill (1802-1870) and Robert Adamson (1821-1848), influenced by painting, tried to "bring out the essential features of any person presented [with] dignity and depth of their perception" (Kracauer 1980:260), attempting to decipher a difficult-to-understand text.

Similar comments have been made by several famous photographers, for example those by Henri Cartier-Bresson (1908-2004):

> I went to Marseille. A small allowance allowed me to get along, and I worked with pleasure. I had just discovered the Leica. It has become an extension of my eye, and I have never been separated from it since I found it. I prowled the streets all day, feeling very strung-up, and ready to pounce, determined to "trap" life – to preserve life in the act of living. Above all, I craved to seize the whole essence, in the confines of one single photograph, of some situation that was in the process of unrolling itself before my eyes (in Sontag 1973:291).

Cartier-Bresson is always associated with expression "The decisive moment", the possibility of capturing an event that is ephemeral and spontaneous, in which the image represents the essence of the event. The skill of the photographer will be crucial. Their eye must see a composition or an expression that life itself offers, and they must know with intuition when to click the camera (in Berger 2013).

In addition to "essence", many comments on photography use words like "magic", "soul", and "revelation". The target of the French photographer Robert Doisneau (1912-1994) is "an isolated image whose contents possess the magic power of remembrance or memorialisation" (Hill and Cooper (1992). The American photographer Edward Weston (1886-1958) notes that, if all the elements combine, there is the possibility of achieving that moment of epiphany photography may surpass other arts:

> Photography's great difficulty lies in the necessary coincidence of the sitter's revealment, the photographer's realization, the camera's readiness. But when these elements do coincide, portraits in any other medium, sculpture or painting, are cold dead things in comparison (in Dyer 2007).

Under the right conditions, helping the subject to feel at ease, with the right skills, and the correct use of the camera, the photographer could portray this essence of the subject. John Szarkowski (1925-2007) mentions the photographer Holgrave, a character in the novel of American author Nathaniel Hawthorne (1804-1864), *The House of Seven Gables* (1851), saying that sunlight will help him convey the subject's deep character with truth that no painter could, and after several attempts he will manage to unravel the normal appearance of "benevolence, openness of heart, sunny good humor [to show] a man, sly, subtle, hard, imperious, and withal, cold as ice" (Hawthorne Ch. 6, in Szarkowski 2007).

For the American photographer and photography professor Minor White (1908-1976), this "essence" or "presence" even acquires a religious element. "That presence is something sacred, it's our Creator, or it's another force. It's grace" (in Hill and Cooper 1992). "I'm trying to be in contact with our Creator when I photograph. I know perfectly well it's not possible to do it all the time, but there can be moments" (in Hill and Cooper 1992). The American photographer Sheila Metzner (1939-) puts herself in the position of a god:

> This image, trapped in my trap, my box of darkness, can live. It is eternal, immortal. The child in the picture will not age like the living child. It is magic.

> Who, then, having this power to stop time, as a god does, to create immortality, to arrest a moment in life, what is my responsibility? What am I looking for? What do I want to preserve, worship, retain, present as my legacy to the world at large, to myself, my family, my friends? (in Johnson 1995:116)

Steve McCurry (1950-), American photographer, whose most famous photo is the Afghan girl on the cover of the National Geographic, emphasizes that "Most of my photos are grounded in people, I look for the unguarded moment, the essential soul peeking out, experience etched on a person's face (McCurry: azquotes.com).

And for Edward Steichen (1879-1973), pioneer of fashion photography in the United States, photography is sublime and can reach the essence of life on earth: "Photography records the gamut of feelings written on the human face, the beauty of the earth and skies that man has inherited, and the wealth and confusion man has created. It is a major force in explaining man to man" (Steichen 1961).

Diane Arbus (1923-1971), an American photographer specializing in portraying people on the margins of society, comments on the magic and power of the camera: "There's a kind of power thing about the camera. I mean everyone knows you've got some edge. You're carrying some magic which does something to them. It fixes them in a way" (Arbus:azquotes. com).

Barthes mentions the term *satori* (悟り) (1981:49), a Japanese Buddhist term for awakening, understanding and understanding, and in the tradition of Zen Buddhism, satori refers to *kenshō*, "seeing true nature" (Ken = "seeing", and shō = "nature", "essence" or "enlightenment").

The concept of magic is part of philosopher Vilém Flusser's argument in *Towards a Philosophy of Photography*. Flusser emphasizes the magical qualities of images in traditional cultures. But the magic of photography is a new, posthistorical magic, "conjuring tricks with abstractions" (Flusser 2000). Whereas prehistoric magic is a ritualization of myths, modern magic is a ritualization of models that are "programmes", including computer programmes, written by "officials", people who work as a function of an apparatus or computer. These technical and superficial images enchant and attract like traditional myths and magically free the recipients from the need to think conceptually with a second-order imagination (Flusser 2000:16-17). Flusser's conclusion is that the photographer should break this superficiality, "outwit the camera's rigidity [...], smuggle human intentions into its program that are not predicted by it [...], force the camera to create the unpredictable, the improbable, the informative" (Flusser 2000:80).

Is there a way in which a photograph can achieve the "essence" of a subject? The above quotes show that it may; photographers themselves believe that photography, in many cases their own photographs, can reach this essence. In Chapter 5 I conduct research with a group of postgraduate students at the Faculty of Philosophy, Letters and Human Sciences of the University of São Paulo (FFLCH-USP) about two photos: the interpreter-guide of the Scottish photographer John Thomson (1837-1921) on the Yangtze in 1870 and the Young Well-dressed Indigenous Interpreter , who worked for the Brazilian army officer, Marechal Rondon, to try to find out if it is possible to agree on a possible "essence" of their characters. Will this "essence" be available to everyone, or will each person see different things? The American photographer Sally Mann mentions Samuel Beckett's play *Endgame*, in which Hamm tells the story of visiting a madman in his cell. Hamm took him to see the beauty of the wheat growing and the fleet of herring fishing boats, but all the madman saw was ashes (in Johnson 1995:134).

3. The Density and the Multiplicity of the Photograph

On the other hand, the photograph's density, complexity and intertextuality are mentioned by several critics. For Susan Sontag (1933-2001):

> The camera makes reality atomic, manageable, and opaque. It is a view of the world which denies interconnectedness, continuity, but which confers on each moment the character of a mystery. Any photograph has multiple meanings; indeed, to see something in the form of a photograph is to encounter a potential object of fascination. The ultimate wisdom of the photographic image is to say: "There is the surface. Now think – or rather feel, intuit – what is beyond it, what the reality must be like if it looks this way". Photographs, which cannot themselves explain anything, are in exhaustible invitations to deductions, speculation, and fantasy (Sontag 1973:38-39).

This complexity, for Sontag, what lies beneath the surface of a photo, will be one of the central themes of this book. For the Franco-Hungarian photographer Brassai (1899-1984) "Photography reflects the infinite variety of subject matter offered by the natural universe" (in Hill and Cooper 1992). In *The Photograph* (1997) Graham Clarke (1941-2007) says that photography develops a "complex intertextuality", to which the German artist Victor Burgin (1941-) refers. In the dense text of photography, ideology constructs meanings and stamps its own power and authority, but it would also be possible to read other codes and texts, and photography can be read as a series of complex and ambiguous simultaneities, containing the codes, values and beliefs, often conflicting, of various aspects of culture and inserting different discourses and references:

> [...] in which is situated not so much a mirror of the world as our way with that world; what Diane Arbus called "the endlessly seductive puzzle of sight". The photographic image contains a "photographic message" as part of a "practice of signification" which reflects the codes, values and beliefs of the culture as a whole [...] Far from being a "mirror", the photograph is one of the most complex and most problematic forms of representation. Its ordinariness belies its ambivalence and implicit difficulty as a means of representation" (in Clarke 1997:28-29).

Brassai, the specialist in photographing the city of Paris, states that "What attracts the photographer is precisely the chance to penetrate inside phenomena, to uncover forms... He pursues them into their last refuges and surprises them at their most positive, their most material and true" (Brassai:azquotes.com).

Siegfried Kracauer emphasizes the tendency of photographs towards the disorganized and diffuse that marks them as records. Therefore, it is inevitable that they are surrounded by a fringe of multiple and indistinct meanings.

So the photographer will, often unconsciously, capture these subtleties and the multiple strands of life. For the pioneer of American photography, Alfred Stieglitz (1864-1946), there is a reality so subtle in photography that it becomes more real than actual reality (Stieglitz:azquotes.com). Vilém Flusser emphasizes the photograph's many points of view; it is thus hostile to ideology and the insistence on a single perfect point of view, thereby introducing "a phenomenological doubt". When a problem arises, photographers

> discover that the viewpoint they have adopted is concentrated on the "object" and that the camera offers any number of different viewpoints. They discover the multiplicity and equality of viewpoints in relation to their "object". They discover that it is not a matter of adopting a perfect viewpoint but of realizing as many viewpoints as possible (Flusser 2000:38).

Photography is therefore complex; it has multiple connections and intertexts; a photograph can be full of subtleties, and it can penetrate phenomena and discover new forms. This complexity can be seen in a number of the photographs examined, which come from a variety of backgrounds, some of which were unfamiliar to me. In Chapter 1, we will see the complex network of meanings and discourses present in the relationship of the diplomat and sinologist Arnold Vissière with China and Chinese diplomats after the Sino-French War of 1884 to 1885 and, then, in the official visit to Paris in 1910 of the Chinese delegation. We will also see the friendship between diplomats, the beginnings of French aviation, its glamour, and the commercial competition between French aviation companies, clothing habits in France in 1910, especially in relation to men's hats, a young 12-year-old boy at work, chuchotage interpretation and, last but not least, the biographies of the main actors.

Likewise, in Chapter 2, we can point to this complexity that runs through the photographs: Jewish immigration to the United States and the history of Jews in the West; relations between capitalism and indigenous groups; the settlement of the state of Nebraska; relations and wars between the various indigenous groups, especially the Pawnee and Sioux-Lakota; the interpreters of the indigenous languages themselves; the history of the photographic studio, and the way in which the Washington DC government managed to take a good part of the lands of the indigenous groups.

Chapter 3 brings us to Brazil and the history of the SPI, the Indian Protection Service, the forerunner of FUNAI, the National Foundation of Indigenous Peoples; we discuss the corruption of the organization, the history and characteristics of various indigenous groups, especially the Xetá and Kaingang from the southern Brazilian state of Paraná, the Xavante and Ariti from Mato Grosso, and the Ka'apor from the border between Pará and Maranhão, where we meet the anthropologist Darcy Ribeiro on one of his first expeditions for the SPI, and then we follow the career of Major Libânio, from the Iriti group, guide-interpreter of Marechal Rondon, an important supporter of indigenous peoples in the early 20th century.

Chapter 4 shows early contacts with the Korubo group in the Amazon and continues through the 2012 Rio+20 Earth Summit, where, initially through the lens of a snapshooter, we meet Raoni and Megaron, important fighters for indigenous rights and forest preservation.

In Chapter 5, we follow the first photographic expedition into the interior of China and attempt to trace Marechal Rondon's indigenous interpreter, portrayed in a 1914 photograph, and initially found on Reddit, the social media site.

In Chapter 6, in the "Devil's Paradise", on the banks of the Putumayo River on the border of Peru with Colombia and Brazil, we confront the tragedy of the massacre of indigenous people who were rubber extractors working for the Casa Arana company. We follow the testimony of John Brown, interpreter for the explorer Captain Thomas Whiffen, and a witness in the report to the British government of Sir Roger Casement, who would die tragically in 1916, hanged by the British government for treason for having imported arms from Germany and exported them to Ireland to help the Irish separatist movement.

Chapter 7 takes us into post-World War II US policy and international diplomacy during this period. We see the policy of US domination in its "backyard", Latin America, and the resistance on the part of many Latin Americans, and especially the events in Caracas on 13 May 1958 when interpreter Colonel Vernon Walters and Vice President Richard Nixon managed to escape from a very dangerous situation, which, if it had gone the wrong way, might have led to an invasion of Venezuela by the United StatesWe also see the consequences of an army officer being the official interpreter for Nixon, the relationship between Walters and the "official" interpreters of the White House State Department, and the careers of Walters and Nixon as, upon his return to the United States, Nixon received a hero's welcome, an event that also helped his career. Likewise, in the photograph of Walters with General Castelo Branco, *Photograph 3: Friends and Brothers,* there are several intertexts: the relationship between the United States and Brazil; the domination of Uncle Sam in Latin America; the role of the USA and Walters in the 1964 military coup in Brazil; and the personal friendship between Walters and Castelo Branco.

In researching Walters' "unofficial" life, we see connections with organizations, some Catholic, of the extreme right, and his possible involvement in various classified actions in the United States.

In the final chapter, Chapter 8, we accompany the career of the Soviet interpreter Viktor Sukhodrev. We follow his childhood in London during World War II and then his career as an interpreter for Soviet leaders Khrushchev, Brezhnev, Gromyko, Kosygin, and Gorbachev. We discover details of the lives of Khrushchev and Brezhnev, and we meet Richard Nixon again, now President of the United States. We witness several summit meetings, behind-the-scenes problems, and the relationship between Sukhodrev and Soviet and American leaders.

Geographically the study is wide. In Chapter 1, we start in China with Arnold Vissière, and then we meet up with him again in Paris. In Chapter 2 we are with Julius Meyer in the state of Nebraska in the United States. In Chapter 3 we arrive in Brazil, following the SPI in Paraná, Mato Grosso, and Maranhão. Chapter 4 takes us to the Amazon and Rio de Janeiro. In Chapter 5 we travel with photographer John Thomson to the interior of China, but we also discuss Portuguese interpreters in Africa and Macao and visit the Brazilian states of Amapá and Mato Grosso. Chapter 6 places us in Peru, on the border of Colombia and Brazil. In Chapter 7, we follow the US army officer Vernon Walters on his secret missions in many countries, but we detail his performance in the 1964 Military Coup in Brazil and analyze photos showing him interpreting Vice President Richard Nixon in Venezuela. In the final chapter, Chapter 8, we follow the interpreter Viktor Sukhodrev, initially in his childhood in London, and later in Moscow, and on several visits to the United States.

We also see a great diversity of types of photograph. In Chapter 1 we examine studio and newspaper photos. In Chapter 2 the photos are of outdoor and studio groups, and one of them is a staged scene. In Chapter 3 we analyze official photographs from the SPI collection, and in Chapter 4 a variety of photos: an "artistic" photo by the contemporary Brazilian professional photographer Ricardo Beliel, a snapshot, and a journalistic photograph published on the Internet. The initial photos in Chapter 5 are by John Thomson (1837-1921), who would become one of the most famous British photographers, and then we analyze a 1914 Brazilian studio photo. In Chapter 6, we have a series of photos of John Brown, two photos taken by the Portuguese photographer Silvino Santos (1886-1970), a book cover, and four other snapshots. In Chapter 7 we use two *LIFE* magazine photographs, and a Brazilian press photograph. Finally, in Chapter 8, we have a variety of official photos of Viktor Sukhodrev, one of them from *LIFE*. In addition, we have an amateur photo taken by Walter Holloway, Uncle Jack, Sukhodrev's London neighbour and friend.

However, most of the preparation of the book was carried out during the Covid pandemic, and all the photos were found on the Internet. Never was I actually able to touch and handle any physical photograph. Elizabeth Edwards and Janice Hart discuss the very different experiences and understandings of looking at the same image on a computer screen and as an albumen print pasted in an album as "the 'grammar' of both images and things is complex and shifting" (2004:2). I am reminded of my sorting through bags and boxes of photographs of my recently deceased mother and my intention of putting the best ones online: two completely different viewing experiences. Photographs can take on many material forms: "Postcards, albums, campaign buttons, decorated photographs carried behind the open coffin at Russian funerals, the photographic placards of mass demonstration or political processions, T-shirts or photographs of ancestors worn in the dance in a Native American community (2004:12). I wasn't able to handle the coffee table book *La Chine: une passion française*; touch Julius Meyer's carte de visite (Chapter 2); see John Brown's wrinkles on the original cover of *Colombia's Forgotten Frontier: A Literary Geography of the Putamayo*; and browse through the paper issue of *LIFE* with Viktor Sukhodrev on the cover (Chapter 8).

4. The Instant, Memory, History and Stories

For Siegfried Kracauer a photograph creates a fixed moment in time as a memory, an impression of an important moment; it is not tied to a specific time and does not have a specific purpose. The photograph captures the physicality of that moment but removes any depth that might be associated with the memory. Thus, a kind of artefact, the photography itself, is created (in Leslie 2010:123-135).

For the British art critic and novelist John Berger (1926-2016), memory implies a certain act of redemption. What is remembered was saved out of thin air. And what is forgotten has been abandoned (Berger 2013:59).

Berger emphasizes the way in which the words of Bertolt Brecht (1898-1956), referring to the actor's art, reflect this importance of the moment:

> So you should simply make the instant
> Stand out, without in the process hiding
> What you are making it stand out from[1] (Berger 2013).

Berger emphasizes the momentary and the discontinuity of photography:

> A photograph arrests the flow of time in which the event photographed once existed. All photographs are of the past, yet in them an instant of the past is arrested so that, unlike a lived past, it can never lead to the present. Every photograph presents us with two messages: a message concerning the event photographed and another concerning a shock of discontinuity (Berger 2013:62).

And for Henri Cartier-Bresson: "As time passes by and you look at portraits, the people come back to you like a silent echo. A photograph is a vestige of a face, a face in transit. Photography has something to do with death. It's a trace" (PhotoQuotes.com).

The photograph is a *memento mori*, which has ghostly qualities:

> Unlike cinema, the photograph holds this recorded moment in stillness, capturing and offering up for contemplation a trace of something lost, lending it a ghostly quality. In this sense, the photograph confronts us with the fleeting nature of our world and reminds of our mortality (Kuhn and McAllister 2008:1).

Photography digs up and brings to light forgotten stories and lives on the borders of our culture and integrates them into new ways of being understood, with the possibility of including them in political action. In the words of Annette Kuhn:

> Memory work is a method and practice of unearthing and making public untold stories, stories of "lives lived out on the borderlands, lives for which the central interpretive devices of our culture don't quite work" [...] As an

---

[1] Bertolt Brecht, "Portrayal of Past and Present in One", from Bertolt Brecht: Plays, Poetry and Prose, Poems 1913-1956, ed. John Willett and Ralph Manheim with the co-operation of Erich Fried, 307.

aid to radicalised remembering, memory work can create new understandings of both past and present, while refusing a nostalgia that embalms the past in a perfect irretrievable moment. Engaging as it does the psychic and the social, memory work bridges the divide between inner and outer worlds. It demonstrates that political action need not be undertaken at the cost of the inner life, nor that attention to matters of the psyche necessarily entails a retreat from the world of collective action (Kuhn 2002:8; and Kuhn and McAllister 2008:14; quote from Steedman 1986:5).

But nostalgia is also an element of memory and photographs. Siegfried Kracauer cites the melancholy that Beaumont Newhall (1908-1993) feels in *History of Photography from 1839 to the Present Day* in Charles Marville's (1813-1879) photographs of the soon-to-be-broken streets of Paris, and Eugène Atget's street scenes (1857-1927) that contain "the melancholy that a good photograph can so powerfully evoke" (Kracauer 1980). The spectator gets lost and identifies with the photo, capturing the estrangement they themself feel. Using the term of T. S. Eliot, the photo is a kind of objective correlative, an image that reflects the viewer's own feelings.

For Kracauer, the function of a photograph changes with the passage of time. After two or three generations, album photos take on documentary functions rather than memory aids: "The grandmother will re-experience her honeymoon, while the children will curiously study bizarre gondolas, obsolete fashions, and old young faces they never saw" (Kracauer 1980). The photos I myself see of my great-grandparents and even grandparents on my father's side are exactly that: generations I never met, and the photos I see of them are empty of any personal element.

This book is an act of remembering, of taking from oblivion the lives and careers of the interpreters Arnold Vissière (Chapter 1); Julius Meyer (Chapter 2); the boy interpreters of the Xetá, Tuca and Kaiuá (Chapter 3); the Xavante interpreter Euvaldo Gomes and Darcy Ribeiro's interpreter, João Carvalho (Chapter 4); Chang (Chapter 5); and even Vernon Walters, a key actor in the 1964 military coup in Brazil (Chapter 7). When I asked students and friends what they knew about Walters, few had any idea who he was and his importance.

In several cases we have information about fragments of their lives. This is the case with Arnold Vissière in Chapter 1, and with Julius Meyer in Chapter 2, and in Chapter 6 we try to unravel the threads of the various biographies of John Brown. In Chapter 7 we know a lot about Walters' official life, but there is much pertaining to state secrets that hasn't been revealed. Similarly, in Chapter 8 we know a great deal about Viktor Sukhodrev's life, but we know little about his formative years between the ages of six and twelve in London, and nothing about the state secrets that were an integral part of his life. In other cases, we have almost no information at all, as with Chang and the Young Well-dressed Indigenous Interpreter in Chapter 5.

We use our imagination to invent stories for them, based on the information and clues we have in the photographs to fill in the gaps. We ponder over Vissière's passion for China and the Chinese in Chapter 1. In Chapter 2 what was the reason for Meyer's gunshot death in Hanscom Park, Omaha in 1909, under highly mysterious circumstances? Was it suicide, something to do with money, a love affair, or perhaps murder? Was there any connection to the links he had with the various Indian tribes? To what extent was John Brown involved in the torture and killings of the Indians in the Devil's Paradise in Chapter 6? We get involved personally with their lives as we shed a tear for Tuca, Kaiuá, and the lost Xetá tribe in Chapter 3 and are touched as the young Viktor improves his English by accompanying Uncle Jack on his postman's rounds in Chapter 8.

Humans have limited memory. We automatically delete data that is no longer needed. I forget, like all of us, people, books, events, places that are no longer interesting to me. This book is about the forgotten, for the most part, and it is a gesture to bring these human beings out of oblivion.

But there are certain photographs that we will always remember. In *Vietnam War Photography as a Locus of Identity* and *Locating Memory* Patrick Hagopian emphasizes that photographs can be more powerful than any other medium. Despite being bombarded by feature-oriented and documentary films, "Yet it is the still photographs that most people remember" (Hagopian 2008:210). And Hagopian exemplifies with the famous photos of the war in Vietnam: the photograph of the Vietnamese girl running naked after the napalm bomb fell near her, and that of General Loan killing a Viet Cong suspect. These photos had a huge effect on the American public: "Senator Robert Kennedy said 'The photograph of the execution was on front pages all around the world – leading our best and oldest friends to ask… what has happened to America?'" (Hagopian 2008:212-213).

Steve McCurry argues that: "A still photograph is something which you can always go back to. You can put it on your wall and look at it again and again. Because it is that frozen moment. I think it tends to burn into your psyche. It becomes ingrained in your mind. A powerful picture becomes iconic of a place or a time or a situation" (McCurry:azquotes.com).

In this book there is no moment that shows anything similar to the violence of the Vietnam War, but there are certain tense, important and crucial moments, especially in Brazilian culture and history: in *Photograph 2* in Chapter 3, *The Interpreter at Work*, we see Tuca, a nine-year-old Xetá boy, helping anthropologist Loureiro Fernandes to identify an artefact found in the rubble of a Xetá village. Tuca is one of the last surviving Xetá, and without him this information about the Xetá culture would be lost forever. *Photograph 3* in Chapter 4, *A Tense Handshake*, shows one of the first contacts with the Xavante in 1949 after attempts at contact resulted in the deaths of most of those who tried to make contact. *Photograph 1: The First Contact with Civilization*, also in Chapter 4, shows the moment of the first friendly contact with the Korubo group in the Amazon in 1996. *Photograph 1: Nixon Speaks*, and *Photograph 2: Walters Interprets Nixon*, in Chapter 7, show Vernon Walters interpreting Vice President Richard Nixon at the US Embassy in Caracas, Venezuela, in May 1958. The initial impression is of a rather routine photograph; however, the photo was taken shortly after an incident that might have had a profound effect on US-Venezuela relations. I shall not give a spoiler her – go directly to Chapter 7! And at the summit meetings between Soviet leaders and US presidents, where most of the photos in Chapter 8 were taken, with Viktor Sukhodrev interpreting, the leaders were discussing matters vital to maintaining world peace, especially the Strategic Arms Limitation Treaty (SALT).

5. Visual Grammar

For Henri Cartier-Bresson a visual grammar is necessary to understand photography in the same way that knowledge of verbal and written grammar is necessary to understand a language: "What reinforces the content of a photograph is the sense of rhythm, the relationship between shapes and values. To quote Victor Hugo: 'Form is the essence brought to the surface'" (in Hill and Cooper 1992). Several of the photographs reproduced here are analyzed using a visual grammar, following the concepts of Erwin Panofsky (1951), Rudolf Arnheim (1974 and 1988), Gunter Kress and Theo van Leeuwen (2006), and the semiotics of Philippe Dubois (2019). The important elements that will be detailed in the analysis of the photos are the vectors representing verbal or non-verbal communication within the photo; the importance of left and right; and the positioning at the top, bottom and centre and in the foreground and background, and outside the photo at the vanishing point; the use of light; the depth of the image; and the angle at which the photo was taken: frontal, oblique or horizontal. And just as the poet uses words and language, perhaps breaking the rules of grammar to create their poetry, the photographer uses the visual elements and grammar of the image to create a poetic image. For Robert Doisneau "The choice of words, the bouquet of words without logical construction, is the same as that within a photo" (in Hill and Cooper 1992).

I try to integrate the formal visual examination with other methods of analysis: the history based on news from contemporary newspapers and magazines, which helps to give a strong narrative element, as in the case of the visit of the Chinese delegation to Paris in 1910 in Chapter 1 and of the visit of Walters and Nixon to Venezuela in Chapter 7; the narrative of my own discovery of who the Indians were in *Photograph 5* of Chapter 4; the biographical, as in the life story of Julius Meyer in Chapter 2, John Brown in Chapter 6, and Viktor Sukhodrev in Chapter 8; the concept of "performance" in daily life, based on the work of Erving Goffman (1961 and 1990), especially important in the case of Julius Meyer in Chapter 2; the use of questionnaires to elicit opinions from other people about two photos, Chang and the Young Well-dressed Indigenous Interpreter in Chapter 5; and, in the absence of data, I resort to a speculative element, as in the case of the photo of Vissière with the Chinese diplomat Luo Fenglu, Chang and the Young Well-dressed Indigenous Interpreter in Chapter 5.

Likewise, I draw on a variety of sources, in addition to "traditional" academic books and articles, websites, documentaries, journals, autobiographies and biographies, interviews, popular magazines, especially *LIFE*, and of course, a few of the millions (or billions) of photos circulating on the Web.

Thus, the sources of each chapter were quite varied. I found little material about Arnold Vissière in Chapter 1, but it was possible to read the reports about the visit of the French delegation to the airfield of Issy-les-Moulineaux in May 1910 in contemporary French newspapers, on the website of the Bibliothèque nationale de France. There was a little more material about Julius Meyer in Chapter 2, including a fictionalized biography, but his mysterious death has never been resolved. There is a lot of material available on the SPI, the Indian Protection Service, the subject of Chapter 3, and the archives of the Museu do Índio were very helpful. In Chapter 4, there is a lot of information about the Korubo on the Internet, and Raoni and Megaron are well-known figures in Brazil. In Chapter 5 the only information we have about Chang is what photographer John Thomson wrote about him, and it was difficult to obtain information about the Young Well-dressed Indigenous Interpreter. I was surprised by the number of publications on the Putumayo in Chapter 6, several of them on Sir Roger

Casement. Vernon Walters, subject of Chapter 7, played a central role in the 1964 coup in Brazil, but I was surprised to find no articles or theses about him as an interpreter, and there must be much classified material about Walters in the Pentagon archives. And, with the help of Marina Darmaros and Google Translate, it was possible to consult material in Russian for Chapter 8. However, the discreet Viktor Sukhodrev never revealed any state secrets.

6. Photographs of Interpreters

Several works published in recent years in Translation Studies, or rather Interpretation Studies, deal with memory and photographs. We can even say that there has been a certain visual turn in the discipline. One of the most interesting studies is the article "Interpreters, photography and memory: Rabinovitch's private archive" (2016a), by Jesús Baigorri-Jalón, which analyzes several photographs from the history of the Rabinovitch family of interpreters, focusing mainly on Georges, who, in 1947, became Interpreter-in-Chief of the United Nations (UN), through an interview with Joana Rabinovitch, daughter of Georges. Baigorri-Jalón manages to extract information by asking about the photos in his private archive. describes his work as a micro-history study, admitting his debt to Carlo Guinsburg, and concludes that it "adds new geographical and institutional elements to our mapping of the sociological development of conference interpreting, including its incipient feminization, in the first half of the 20th century, an extended chapter of a larger history, the evolution of the interpreting profession microcosm" (Baigorri-Jalón 2016a:187).

In "The use of photographs as historical sources, a case study" (2016a) Baigorri-Jalón uses photographs to revive the institutional memory of the UN's shift from consecutive to simultaneous interpretation. The images used are of the interpreters and the audience using the new equipment and the equipment itself. The photos are reminiscent of the change that brought about a new working environment and professional regulations that continue to this day (2016b).

Framing the Interpreter: Towards a visual perspective, edited by Anxo Fernández-Ocampo and Michaela Wolf (2014), recalls unknown performers in times of war and conflict, several of whom, seen as traitors to their native society, would die shortly after being photographed.

Differently to *Framing the Interpreter*, the sociopolitical situations of my study are quite varied. Initially, in Chapter 1, we have the contact between two high-ranking diplomats after the Sino-French War of 1884-1885, then the Chinese military mission in Paris in 1910, which seems to have been successful, at least by the smiling faces we see in the pictures. In Chapter 2, the positive interaction between Nebraska Indians in Julius Meyer's photographs of the 1870s may be a little more questionable if we consider that the introduction of Indian groups into the capitalist system was negative; however, I believe this contact was inevitable and Meyer helped them integrate into the system.

Chapters 3, 4, 5, and 6 deal with photos of indigenous language interpreters in South America, especially in Brazil. Chapter 3 examines photographs of the Indian Protection Service (1910-1967) in Brazil, which was often criticized for its corruption and links with landowners, showing situations of both isolation and integration of interpreters. The interpreter who is closest to the interpreters we saw in *Framing the Interpreter*, rejected by their native society and unable to be part of the society they are serving, is John Brown, subject of Chapter 6, a black American who pretended to be from Barbados and who was the explorer-guide interpreter of Captain Thomas Whiffen.

The power of the interpreter is the theme of Chapter 7, which describes the interpreter and soldier Vernon Walters as a symbol of US power in Latin America, especially his role in the 1964 coup in Brazil, and the analyzed photos demonstrate elements of this domination.

Chapter 8 takes a very favourable view of the Soviet interpreter, Viktor Sukhodrev, whose interpretation often gave a more positive impression of Khrushchev's lack of sophistication and Brezhnev's boorishness. However, he did not use this modifying technique when translating "My vas pokhoronim!" into English at a meeting in Poland in 1956. Sukhodrev's translation of "We will bury you" shocked many people in the West and it was thought that nuclear war might be close at hand.

Luckily, it wasn't!

CHAPTER 1

THE DIPLOMAT –
ARNOLD VISSIÈRE

1. La Chine: Une Passion Française

Arnold Jacques Antoine Vissière (微席葉) (1858-1930), a sinologist who was about to start working at the National Library in Paris, was sent to China by the Ministry of Foreign Affairs to work as an interpreter for the French delegation in Beijing. He began his career as an interpreter in 1880 when he was delegated to the Brazilian mission that was making a commercial treaty with China, and, naturally, French was the lingua franca. He later participated in the peace negotiations after the 1884 -1885 Sino-French Indochina War, which resulted in the 1885 Treaty of Tianjin, allowing France to not only annex the Tonkin and Annam regions to French Indochina, the path to the total colonization of what is now Vietnam by France but also to guarantee the opening of eleven new ports to the West, bringing freedom of movement for European merchants and Christian missionaries through various commercial treaties between France and China.

In 1899 Vissière accepted the chair of Chinese at the School of Oriental Languages in Paris, where he developed a method of transcribing Chinese into French, trained future interpreters, and wrote a large number of works on China and the Chinese language. He continued to play an active role in the Ministry, receiving Chinese delegations in France, and was appointed Secretary Interpreter of the Ministry in October 1899, Consul General in January 1907, and Minister Plenipotentiary in February 1916, thus combining his academic work with the responsibility of receiving diplomatic missions from China. Writing in 1899, the sinologist Edouard Chavannes declared that Vissière "was undoubtedly the man in France who had the most profound knowledge of the spoken Chinese language and their way of doing business"[2] (in (Gaspardone 1930:650).

Ruth Roland stresses the importance that France gave to its foreign delegations. Napoleon created a school of oriental languages within the Ministry of Foreign Affairs. Exams for joining the Foreign Service were difficult, and Roland emphasizes the French belief that "a diplomat ought to be a representative of the *best* that his country has to offer" (Roland 1999:139-140).

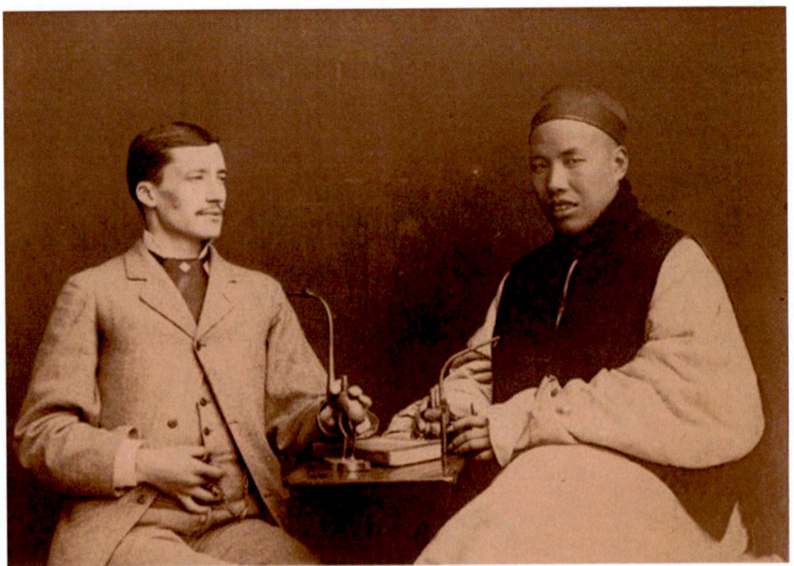

Photograph 1: La Chine: Une Passion Française

Luo Fenglu (羅豐祿) (1850-1903) entered the Fujian Arsenal Academy in 1867, and as the best student in 1877 he was elected to be one of the first students sent by the Qing dynasty to study in Europe. Luo arrived at King's College London in May 1877. During his stay in the UK he also worked as a translator for the Chinese embassies in the UK and Germany. He was also naval secretary to Li Hong Zang (李鴻章) from 1880. Sarah Paine gives us an interesting description of Luo.

---

[2] "qu'il était incontestablement l'homme de France qui a la connaissance la plus approfondie de la langue chinoise parlée et du style des affaires".

During the peace negotiations of the 1894-5 Sino-Japanese War at Shimonoseki, Premier Ito Hirobumi had questioned one of the Chinese translators, Luo Fenglu, a brilliant man who had been educated in the West and had a broad knowledge of western literature. Premier Ito wondered why China had not learned more from the West, to which Luo responded, "You see, in our younger days we knew each other as fellow students, and now you are Prime Minister in your country and I am an interpreter in mine". In Japan, those with foreign expertise rose to the top while in China such persons were relegated to inferior position (Paine 2002:301).

However, it appears that Luo's superiors were aware of this as he later became the Chinese ambassador to the United Kingdom (1897-1901).

Photograph 1 was used on the cover of the book *La Chine: une passion française* by Isabelle Nathan-Ebrard, published in 2014. It is a richly illustrated book based on the French diplomatic archives in China. This title gives us some clues about the meanings we can extract from the photograph, dated 1884.

Despite the hostility between France and China during the Sino-French War of 1884-1885, the photo conveys tranquility and affection. We can assume that the meeting has some connection with the war, but it seems more like a meeting between friends, interpreters and colleagues smoking together in a convivial relationship. Traditionally, smoking together with other people represents friendship, togetherness, and peace. In the next chapter, we will see the importance of the peace pipe in North American indigenous culture. France and China hope for peace and must develop a positive relationship. The photograph is a *studium*, following Roland Barthes, showing us that, with patience, affection and understanding, harmonious relationships can be developed with the mysterious and inscrutable Other. The photo also demonstrates a desire for the Other, a Europe drawn to the beauty and mystery of the East.

Photograph 1a: La Chine: une passion française - Cover

While the "chinoiserie" fashion was more of a mid-18th century phenomenon, the cult of the Chinese exotic persisted into the 19th century. Jonathan Spence describes these various exotic elements: "One was an appreciation of Chinese grace and delicacy, a sensitivity to timbre and texture, that spread from the initial stimuli of silk and porcelain and temple architecture to become the basis of an entire aesthetic". And this is mixed with a sensuality that includes something more severe and aggressive, something unknowable, dangerous and intoxicating, composed of perfume and sweat, of heat waves and rotting night air. Then there is the sinister element, the "realm of Chinese violence and barbarism, of hidden cruelties, threats of ravishment, uncontrollable impulses". And China was seen "as the realm of melancholy, as a land that stood for something forever lost", lost to the materialism of the West and the weakness of its own state. In addition, Spence mentions opium as a natural accompaniment to such melancholy (Spence 1999:307-308).

This exotic China can also be seen in other works, such as the diary (1864/1985) of army captain Prosper Giquel (1835-1886), the novel of Pierre Loti (1850-1923), *Les Derniers Jours de Pékin* (1901), the prose poems (1957) of Paul Claudel (1869-1955), and the modernist poetry of Victor Segalen (1878-1919) in *Stèles* (1982) and his novels *René Leys* (1974) and *Le Fils du Ciel* (1985). In fact, Vissière had personally encouraged his student Segalen to study Chinese culture.

La Chine: Une Passion Française shows us a clear triangulation, in which initially leads us to look at Vissière, on the left of the photo, as we Western viewers are used to reading from left to right. Vissière, aged twenty-seven, looks captivated and is gazing at Luo Fenglu. We hope that Luo will return the look of affection and admiration; instead, he looks directly into the camera. Indeed, he seems more interested in the camera and the photographer, thus extending the field of the photograph towards and beyond the photographer to the vanishing point outside the photograph. For Philippe Dubois, a photograph can be extended at the front, integrating the photographer as an invisible partner, as in this case, giving the photograph a profundity of image.

> [...] for an out-of-field that works *in the depth of the image*, or rather, *in its advance*, which no longer overflows at the sides, but at the front, for what is in fact at the origin of the cut. An out-of-field, therefore, that positions the operator explicitly, that integrates him more or less as an invisible partner, that designates his place, which is the very place of the constitutive gaze of the scene and of the field itself [3] (Dubois 2012:183).

So, with Luo's gaze directed at us, our eyes are drawn to the centrepiece of the photo, the table, and we wonder what the book is about. And if we

---

[3] "para um fora de campo que trabalho *na profundidade do da imagem,* ou melhor, *em seu avanço,* que não transborda mais pelos lados, mas pelo frente, pelo que está de fato na origem da corte. Um fora de campo, portanto, que posiciona o operador explicitamente, que o integra mais ou menos como parceiro invisível, que designa seu lugar, que é o próprio lugar do olhar constitutivo da cena e do próprio campo".

consider Vissière's gaze to be homoerotic, we also notice that the stem of his pipe is much more erect than Luo's.

Although this interpretation may seem a bit exaggerated, we can certainly find a bonhomie, a mutual sympathy, a meeting of minds, a spirit of the "club". Harold Nicolson, himself a leading British diplomat, describes the "club" of diplomats, albeit from different cultures, saying that such diplomats had similar levels of education, experience and similar aim. They wanted the same kind of world, almost developing a corporate identity independent of their national identity, and believed that the aim of diplomacy was the preservation of peace (Nicolson 1954:75).

In this photo we can emphasize the importance of the look, something that we will see again in several photos in this book. The way in which the photograph can capture "something inside", a *noema*, an essence, emotions such as love, compassion, grief, ardour, desire, even the "mad truth", according to Roland Barthes (2015:93). Isn't this what we see in Vissière's eyes? The passion, his love for China, the tranquility, the affection, the attraction (sexual/cultural/fraternal) that he feels for Luo?

This becomes clear when we compare this photo with another photo we have not been able to include, which shows Vissière with a group of other French interpreters in China. Vissière is frowning, bored, weary of the company of his fellow Frenchmen, and he seems to be taking his watch out of his pocket to see how much time is left for the meeting, and dreaming of China and the Chinese...

To summarize this section, we can use the three levels of Panofsky's image analysis (1939, in Burke 2001:35-36) to examine *Photograph 1*. At the initial level, there is a friendly meeting between the two diplomats after the Sino-French War. At the secondary level, we notice the importance of clothes, looks, pipes and their symbolic value. And in the third, the iconic value is the harmony of Sino-French relations and the great attraction that China had for many French people.

We can complement this with the two "broad categories of history" Elizabeth Edwards mentions in her analysis of anthropological photographs: firstly "the forensic, constituted by the material stability of the content in terms", similar to Panofsky's second category, and then "the submerged, traced through refiguring histories that gather around and enmesh images", again the history of contact between France and China (Edwards 2001:87).

2. What are they looking at?

Photograph 2: What are they looking at?

In the next five photos, from the Rol press agency, we find Vissière again, twenty-six years later, now fifty-two. His girth is a little wider, but he is still immaculately dressed. He has been married since 1889 to Marie Victoire Eugènie Varnier and has four children: Simone (1891-1970), Gaston (1896-1966), Philippe (1917† in the First World War), and Frédéric (1900-1968).

In *Photo 2* we see a Chinese delegation. It is a mixture of civilians and military, along with several Europeans, two of whom, one of them Vissière, are extremely well dressed in top hat and tailcoat. Everyone is looking at something in the distance.The vanishing points of all converge (Arnheim 1974:286-287). What they are looking at?

We are at Issy-les-Moulineaux aerodrome, southwest of Paris, on the morning of Monday 16 May, and so it becomes clear that everyone is looking at a plane in the sky. In fact, as the first flight made in China took place only on 21 February 1911, it may be the first plane in flight ever seen by several members of the Chinese delegation[4]. Prince Tsai Tao (濤濤, Zaitao) (1887-1970), Minister of Military Consultancy, uncle of Puyi (溥儀) (1906-1967), the last Emperor of China, has just arrived from King Edward VIII's funeral in London, which was held on 6 May. He is traveling to Japan, the United States, Great Britain, France, Germany, Italy, Austria and Russia to observe and learn from these countries' most advanced armed forces, along with Count Li Qing-Mai, General Ha Han-Chang, the president's military adviser, second from the right, and other officers. Tsai Tao would be a great survivor of the revolutions in China in the 20th century, becoming a military adviser to the Nationalist government and the People's Republic. He died only in 1970, at the ripe old age of 83.

Photograph 3: Blériot XI

The 17 May 1910 edition of the sports newspaper, *L'Auto*, reports that the visit began with an exhibition of the Ville-de-Bruxelles airship's basket measuring 44 metres in length, and the Chinese mission was particularly impressed by the propellers. At ten o'clock, an Astra balloon was taken out of the hangar onto the runway. Visitors then witnessed a solo flight by ace Alfred Leblanc.

---

[4] Made by the French aviator René Vallon, who took off from a racecourse in Shanghai.

In the photo, Vissière is talking to Tsai Tao and, to the left of the photo, we see, also in topper and tails, Édouard Surcouf (1862-1938), director of the Astra company, who has organized the technical side of the visit on behalf of the Automobile Club of France, which is heavily involved in aviation, together with the Aéro Club de France. Surcouf stands next to engineer and aviator Henry Kapférer (1870-1958), wearing a bowler hat. Both were important personalities in French aviation. Surcouf was an engineer, designer, pilot of airships, and an industrialist. In 1889, he was appointed president of the School of Aeronautics. In 1902 he built his first airship, the Astra I, and from 1904 he began working with industrialist Henry Deutsch de la Meurthe manufacturing airships. Le Ville de Paris (Astra II) suffered a serious accident during its maiden flight in December 1904 but was rebuilt and flew again in 1906. Deutsch de la Meurthe joined Surcouf in 1908 to found Société Astra des constructions aéronautiques. This new company ramped up production, manufacturing the Wright brothers' aircraft under licence and their own models, such as the CM. The company also manufactured airships. Surcouf employed several aeronautical engineers, and Henry Kapférer, who had worked with Deutsch de la Meurthe since 1893, would become the manager of this new company.

3: Chuchotage

Photograph 4 is an excellent example of chuchotage – whispered simultaneous translation – and we see Vissière in the classic chuchotage position, just behind the shoulder of the person he is interpreting for, Tsai Tao, translating a conversation between aviator Louis Blériot and Tsai Tao. Blériot was at the height of his fame, having made the first solo flight across the English Channel on 25 July 1909 in his Blériot XI monoplane, and was an aviation consultant for the French government. He made his fortune from 1897 onwards making automobile headlights and soon began supplying his lamps to Renault and Panhard-Levassor, two of the leading automobile manufacturers of the day. Blériot developed his first models at the Ateliers d'Aviation Edouard Surcouf, Blériot et Voisin, active from 1905 to 1906, but the partners would soon go their own ways. Between 1909 and the outbreak of World War I in 1914, Blériot's company produced around 900 aircraft, most of them variations of the Type XI model. In July 1910, the Bleriot Type XI held world records in the areas of duration, distance, speed, and altitude. Blériot monoplanes and Voisin-type biplanes, together with the Farman derivatives of the latter, dominated the aviation market before the war.

Photograph 4: Chuchotage

The ten-minute circular flight was made by aviator Alfred Leblanc (1869-1921), friend and right-hand man of Blériot, the first pilot in France to qualify through an exam. It was a great summer for Leblanc: he rose to fame and became a celebrity and pin-up boy by winning the Circuit de l'Est race, covering the 805 km in twelve hours, one minute and one second, at an average speed of 66.99 km/h, piloting a Blériot XI with a Gnome engine. In November 1910 he represented France in the Gordon Bennett Trophy air race held in New York, but on the last lap his engine stopped, and he was forced to make a landing, hitting a telegraph pole, but fortunately he did not suffer serious injuries. If the engine failure had not happened, Leblanc would have won the competition instead of Claude Grahame-White, who was also piloting a Blériot. During World War I, Leblanc was the general manager of the Blériot factory in Suresnes.

L'Auto tells us that Leblanc landed without a hitch, despite considerable turbulence, and was greatly applauded. Blériot decided not to take any chances and stayed on the ground. He had had a serious accident the previous December in Istanbul when, flying in gusty weather conditions, his plane came to a crash landing on the roof of a house, leaving him with internal injuries, several broken ribs, and a three week stay in the hospital. Several months later he may still have been feeling the effects (Hoddenbach:Internet).

A number of the photographs in *Framing the Interpreter* emphasize the isolation of the interpreter. In Rachael Langford's "Framing and Masking", Figure 3.2. "My standard bearer and interpreter on a tour through Imperri (1898)"), the interpreter was later killed in a native uprising in the British protectorate of Sierra Leone, and this "is a haunting image of the perils of inhabiting the contact zone". The viewer's gaze and that of the indigenous observer in the background meet and are "opposing gazes, each looking at a different side of the interpreter and metaphorizing the colonial auxiliary's isolation from both sides of the border where colonial power meets opposition to that power" (Langford 2014:44).

Langford uses Mary Louise Pratt's concept of the contact zone: "social spaces where disparate cultures meet, clash, and grapple with each other, often in highly asymmetrical relations of domination and subordination – like colonialism, slavery, or their aftermaths as they are lived out across the globe today" (Pratt 1992:4). Cultures separated by geographic boundaries now intersect and interact, often on the colonial frontier. In the same article, Figures 3.4, postcards of "Dragoman, guide for travellers" and 3.5, "Arab interpreter" highlight the exoticism of the subjects' clothes and weapons, and "What is strikingly absent from the image (3.4) is the notion of the interpreter inhabiting a contact zone" (Langford 2014:48). Neither of the two photographs gives us any idea of any kind of interpreting performed or contact with any other human being.

However, in our photographs the circumstances are very different. Although France conquered Vietnam, which had been dominated by China, this seems to now be part of the past. The photographs are taken by the Rol press agency, and what we see is the interpreter Vissière bringing together French aviators, industrialists, and members of the Chinese diplomatic mission. The "contact zone" in the sequence is of harmony, mutual interest, trade, cooperation, and exchange.

Kress and van Leeuwen examine images as semiotic social signs with "represented participants", people and things shown in the image, and "interactive participants", producers and spectators (2006:114). In this series of photos taken in Issy-les-Moulineaux aerodrome, the interest is mainly the result of the photograph, the "participants represented" – the pleasant visit to Issy of the Chinese delegation – the first time that several of them may have seen a plane in the sky, the short balloon ride of two Chinese army officers, the possible military uses of planes, aided by their interpreter, Vissière, who helped break the ice and make them feel at ease. We, the spectators, enter this visit and glimpse the early days of aviation, a positive side of Sino-French contact, and see the interpreter's successful chuchotage interpreting. On the other hand, *Photo 1*, with Luo Fenglu looking directly at the camera, engages the viewer much more, and we become part of the triangular relationship between Vissière, Luo Fenglu and the camera.

4. The Boy on the Left

Photograph 5: The Boy on the Left

Photo 5 shows Leblanc immediately after his flight being applauded and everyone taking off their hats to greet him. All we can see of Vissière is his top hat, but the photo gives us a good view of the Chinese mission and includes the army officer, General Ha Han-Chang, as well as a working boy, about twelve years old, on the left. Interestingly, this was one of two photos printed in *Dimanche Illustré*, which is much more interested in showing the Chinese mission than the interpreter!

Cropping a photo can distort the image, focus on one element and purposefully crop others. Dubois stresses that photographer always cuts, separates, and omits elements that cannot be seen in the photo. He describes each shot as an axe that excludes, rejects, and denies (Dubois 2012:178). John Berger complements this idea by saying that photography invokes what is outside the photo (Berger 1980:293). For John Szarkowski the photographer's central problem is what he will decide to include and reject, where the edge of the photograph will be. His decision can result in unexpected juxtapositions and, "by surrounding two facts, it creates a relationship" (Szarkowski (2007:70).

So here we have an apparently uncropped photo, and the boy stays inside the published photo. In 1910 the cutting techniques were more time consuming, and the agency's photos had to be ready in a few hours to be printed in the next day's newspapers. There probably would not have been be enough time to cut the boy out. And in 1910 the fact that there was a boy of about twelve years-old working at the airfield would not cause any surprise as there was nothing illegal about this. However, nowadays the situation has changed, and the boy attracts our attention. What was supposed to be a photograph showing the applause that the ace Alfred Leblanc receives after his flight ends is for us today a photo that includes what must remain outside, as Dubois states. Thus, the figure of the boy is highlighted, and he, as a proletarian worker, is contrasted with the wealthy engineers, aviators, businessmen, diplomats, and Chinese dignitaries[5]. And this leads us to the next question: how many other twelve-year-olds are in menial working conditions at the Issy aerodrome?

The appearance of the working boy is, for us, the *punctum* of Barthes. The boy appears isolated on the left of the photo. He is not part of the group around Prince Tsai Tao and stares at Édouard Surcouf and/or General Ha Han-Chang. Behind him we see another boy of a similar age, of a higher social class, who will be one of the foci in the next section. We compare the clothes of the working-class boy, wearing his apron and woollen jumper,

---

[5] I write this on 26 January 2020, when one of the news items of the day is the cropping of the photo of Ugandan climate activist Vanessa Nakate, in which she appears alongside other activists, among them Greta Thunberg (UOL). In the original photo, taken in Davos, Switzerland, Vanessa was the only black woman with four other white women, and the apparent reason for cropping was the building behind Vanessa, which would spoil the photo. We can imagine that our photo could easily be cropped nowadays.

with the suit and high collar of the boy in the bowler hat, and the somewhat dandy straw boater-type hat of the taller teenager behind him.

As in all industrialized countries, French children made up a large percentage of the workforce. In 1840 women and children accounted for 75% of the workforce in the textile industry, and it was not until 1896 in France that the law prohibited any child under the age of twelve from working (Les Droits des Enfants:Internet).

Returning to the main action of the photo, we see that Kapférer is taking off his hat as a sign of respect to the aviator Leblanc, who had just completed his flight. We see that Vissière's top hat is not on his head. The very tall man, probably an engineer who works for Blériot, who appears in other photos and films with Blériot, is also doffing his hat, and it is very likely that others have already taken theirs off. In celebratory situations hats, when all the men wore them, may have been thrown into the air.

Rudolf Arnheim distinguishes between the bottom and the top of a picture. The lower part is always heavy, where there is gravitational weight, while in the upper part there is an attempt to escape from gravity, from earthly life, to climb to the sky and clouds, as in many religious paintings, and as aviators and planes do: "To rise upward means to overcome resistance – it is always a victory" (Arnheim 1974:30). For Kress and van Leeuwen (2006), writing about product commercial texts, the upper part is the "ideal", and the lower part the "real":

> "[…] the upper section visualises 'the promise of the product', the status of glamour it can bestow on its users, or the sensory fulfillment it can bring. The lower section visualises the product itself, providing more or less factual information about it, and telling the readers or users where it can be obtained, or how they can request more information about it, or order it" (Kress and van Leeuwen 2006:186).

Thus, the movement of raising their hats represents, in addition to respect, the aviators' attempt to conquer this new frontier of the sky and, in *Photograph 3*, the look upwards, towards the plane piloted by Leblanc, is also the look towards the future, to go beyond terrestrial limits.

5. Vissière Translates Alfred Leblanc

Source gallica.bnf.fr / Bibliothèque nationale de France

Photograph 6: Vissière Translates Alfred Leblanc

Photograph 6 shows an amusing comment being shared, obviously well translated by Vissière. In it we see both the French and Chinese delegations laughing though Tsai Tao has only a wry smile on his face, belying the report of *Le Petit Parisien*, which sourly claims that "the celestials […] left without showing the least astonishment, without giving a single word of enthusiasm"[6] (*Le Petit Parisien* 1910:1).

The other important participant in *Photographs 5* and *6*, dressed in a tweed suit on the left and introducing Leblanc to Prince Tsai Tao, is Henry Deutsch de la Meurthe (1846-1919), "Europe's Oil King", patron, supporter of aviation and main investor in Surcouf's Astra factory[7], whom we have already seen in *Photograph 3*. He has sponsored several prizes to encourage

---

[6] "[…] les Célestes […] se retirèrent sans manifester le moindre étonnement, sans prononcer un mot de sincère enthousiasme".
[7] He is remembered by the Émile and Louise Deutsch de la Meurthe Foundation, the oldest building of the Cité Internationale Universitaire in Paris. Founded in 1925, it consists of six residential buildings and a central administrative building, which also houses a bell and clock tower.

the development of aviation technologies, including the Grand Prix d'Aviation and the Deutsch de la Meurthe prize of one hundred thousand francs for the first machine to make a round trip from Parc Saint Cloud to the Eiffel Tower, a total of eleven kilometres, in less than thirty minutes. This award was won on 19 October 1901 by Brazilian Alberto Santos-Dumont in the airship he designed. Ironically, Deutsch did not seem to have had much faith in getting into a plane, as he only made his first flight as a passenger in May 1911 when he was taken for a ride in a Blériot monoplane piloted by Alfred Leblanc. On 21 May 1911, French Minister of War Maurice Berteaux was killed when a monoplane piloted by Louis Émile Train crashed at the start of the Paris-Madrid Air Race at Issy after making a forced landing, hitting a group of people, killing Berteaux and injuring Deutsch, French Prime Minister Ernest Monis and his son and others.

Writing about the first crossing of the Atlantic Ocean by Charles Lindbergh in 1927, Bill Bryson (2014:138) describes the enormous emotion and interest in this feat:

> At 10.22 p.m. Paris time – precisely 33 hours, 30 minutes and 29.8 seconds after taking to the air […] the *Spirit of St Louis* touched down on the grassy spaciousness of Le Bourget. In that instant a pulse of joy swept around the Earth. Within minutes the whole of America knew he was safe in Paris. Le Bourget was instantly a scene of exultant pandemonium as tens of thousands of people rushed across the airfield to Lindberg's plane – "a seething howling mass of humanity […] surging towards him from every direction of the compass", in the words of one onlooker.

Let us dig a little deeper into this photo and look at it from various angles. We have already described the enormous enthusiasm for aviation, both as a sport and as a business. We can also see the photo as a formal composition, an act of diplomacy, an iconic image and a portrait of Parisian society in 1910. Following Kress and van Leeuwen, we can examine the vectors, the lines of action in the photo. Almost all vectors converge on Prince Tsai Tao. As head of the Chinese delegation, he is the centre of attention, and Leblanc is speaking to him.

Photograph 6: Vissière Translates Alfred Leblanc: Vectors (1910)

Following Mercedes Gaffron, Rudolf Arnheim emphasizes the importance of the left side in the photo. The spectator subjectively identifies with the left side, and everything on that side has greater importance and is more central to the spectator's consciousness. In the theatre the audience looks first to the left side of the stage. According to Gaffron, this reflects the dominance of the left side of the cerebral cortex. On the other hand, what happens on the right side is more conspicuous and heavier. In other words, we "read", at least in Western societies, from left to right (Arnheim 1974:33-36). Here we enter the photo from the left, together with Deutsch de la Meurthe, who is standing away from the larger group. We look with him at the larger group, centred on Prince Tsai Tao, and we see that everyone is smiling, sharing the joke. We see that the French and the Chinese, two peoples so far apart, can have a friendly relationship, and we are happy to see this. The vectors in the photo, the lines of energy, go from almost everyone's eyes to Tsai Tao.

However, there are two people who are more interested in other things, and by forming a *punctum* they break this harmony. The boy in the bowler hat, possibly the son of the tall engineer, Henry Kapférer, seems to have found something very interesting on the ground. And Kapférer himself seems more interested in the camera and looks straight at us.

The fact that Vissière, the diplomat-interpreter, is the head of the French delegation may seem somewhat unusual to us these days, as we are used to seeing interpreters occupying subordinate positions and never taking centre stage, as in Vissière's case. Surprisingly, at least for me, but perhaps not in terms of the general anonymity of translators and interpreters, Vissière's presence was almost ignored by the newspapers that covered the visit: *Le Temps*, *Le Petit Parisien*, *L'Auto*, *Le Gaulois* and *Le Journal*. The newspaper photos focus on the Chinese delegation: *Le Journal* includes a photo, not reproduced here, of General Ha Han-Chang and Tsai Tao looking up at Leblanc's flight; the Chinese delegation showed a "lively curiosity" ["vive curiosité"]; and the fact that an interpreter was present was mentioned though he remained unnamed. *Le Soleil du Dimanche Illustré* publishes *Photograph 5: The Boy on the Left* and that of the balloon ready to take off. *Le Petit Parisien* has a photo of the two Chinese officers, General Liang Pi and Colonel Sin, in the balloon and about to take off. Tsai Tao remains outside the balloon with other members of the delegation. *Le Temps* and *L'Écho de Paris* claim that visitors were delighted with the visit, and *L'Auto*, which has a similar image to *Photograph 4: Chuchotage*, claims that the two officers were "very moved" ["fort émus"] by their balloon trip.

We can notice the great variety of menswear, particularly in terms of hats[8]. Photography is an important way of documenting how people dressed. The excursion organizers Surcouf and Vissière wear top hats and tailcoats. The well-heeled Deutsch de la Meurthe doesn't need to impress anyone, and he wears a fedora and an English tweed suit. In 1910, we are already in the period when all men wore suits, and, on this semi-formal occasion, their use would be almost mandatory, as would the use of hats. Kapférer and the very tall, nameless engineer wear bowler hats and suits, and the two boys wear similar bowler hats, the younger one with short trousers[9].

---

[8] In this chapter, there is no image containing a woman! However, they were very active in aviation, so it is surprising not to see women in the crowd at Issy-les-Moulineaux aerodrome. Amelia Earhart (1897-1937) and Amy Johnson (1903-1941) are household names. Alfred Leblanc was a friend of Raymonde de Laroche (1882-1919), a former actress who was the first woman in the world to obtain a pilot's licence in 1910. She died in a plane crash.

[9] This fashion lasted a long time. I well remember my first suit, with short trousers, which I received in 1962, in Birmingham, UK, when I was six years old. I hardly

Leblanc is wearing a bomber hat, and Blériot, in *Photo 4*, wears an English flat cap, traditionally associated with the working class but also with sportsmen and often seen in horse race meetings in the UK and Ireland. In the crowd we can see fedoras, flat caps, and straw boaters. The Chinese caps seem to be distinguished according to rank, and we can see four different ones in *Photograph 6*.

And *Photograph 6* leaves many puzzles unsolved. Why are there no women in the picture? What is the boy looking at? Is the relationship between the French and the Chinese as harmonious as it seems? Is Vissière still fascinated by things Chinese? Where is Blériot? Was he too scared to fly with the turbulence? How worried are the Chinese about the precarious position of the Qing dynasty? "The photograph awakens a desire to know that which it cannot show" (Edwards 2001:19).

Elizabeth Edwards proposes the co-existence of the "containing" or "originating" (of who, what, why and when?) context and the "dense context", which emerges through the relations of the photograph, "an awareness that certain phenomena, persons or events accumulate layer upon layer of meaning, perhaps finding themselves at a cross-roads of morphological chains, at the intersection of numerous contexts and actions, or a nodal point where both contemporary and modern preoccupations reflect and enhance each other" (Egmond and Mason 1999:249, in Edwards 2001:109).

Thus, we have a huge variety of elements in the series of photos taken in Issy. John Berger quotes photographer Lee Friedlander, who comments on the amount of information that accidentally appears in his photographs, and includes the quote: "'It's a generous medium, photography,' he concluded drily" (Berger 2013:xvi). For Kracauer this is an endlessness" (1980:263-265). Elizabeth Edwards emphasizes the "rawness" of photographs: "Their unprocessed quality, their randomness, their minute indexicality [...] inherent to the medium itself" and, quoting Christopher Pinney, mentions that historians and anthropologists "are worried by still photography because, lacking the constraining narratives of film, still images contain too many meanings (Pinney 1992:27, in Edwards 2001:5). For Pinney, it is impossible to eliminate this randomness: however carefully the photographer may attempt to arrange the contents of the photograph, "the inability of the lens

---

ever wore it, and the fashion ended soon after. Wearing long trousers was a sign of having entered adolescence, and younger boys always wore short trousers, even in cold weather!

to discriminate will ensure a substrate or margin of excess, a subversive code present in every photographic image that makes it open and available to other readings and uses" (Pinney 2006:3).

We can list the subjects contained in the sequence of photographs we have just analyzed: the beginning of aviation; the danger, excitement, and fear of flying; friendly contacts between France and China; the last years of the Qing empire, which would be overthrown in 1912; competition between aircraft manufacturers in France; attempts to make a trade deal; child labour; the wonder of seeing an airplane flying for the first time in one's life; the lives and careers of Blériot, Surcouf, Deutsch, Leblanc, Vissière, Tsai Tao, among others; the various body and facial expressions; men's clothes in 1910; the variety of hats; and, last but not least, a good example of chuchotage, with Vissière as the link, joining, and interconnecting several of these elements and responsible for the resulting joviality and good humour of the photograph. In other words, we use several semiological systems, each one being a social construction: the language of the clothing of the time, facial expressions, body gestures, and the photographic frame.

So these photographs are very rich. John Berger describes different types of photographs. In general a photograph captures a movement and also cuts that movement (a). The photograph will always include information, which will vary depending on the viewer's relationship to the event. When a man riding a horse in a photo is a stranger, there will be much less information (b) than when the man is the spectator's son (c). And when the event(s) in the photo imply other events and are related to other happenings, as in the illustration below, the circle extends beyond the immediate information dimension.

Berger (1980) emphasizes that if photography is to be incorporated into social and political memory it must be read in a radical way, stating that: "A radial system has to be constructed around the photograph […] so that it may be seen in terms which are simultaneously personal, political, economic, dramatic, everyday and historic" (Berger 1980:63, in Kuhn 2002:8).

The North American photographer specializing in portraying nature, Eliot Porter (1901-1990), contrasts centripetal photography with centrifugal photography: in the former, all the elements of the photo are concentrated in a central point to which the eye is drawn; in this one, like a sunburst, the gaze goes to all corners of the photograph, as in the illustration below, with its multiple points of interest.

Illustration 1: Sunburst

Hats and clothes, the flight, the Chinese diplomatic visit and the attempt to make the Chinese interested in French planes make up the primary or natural content of Panofsky (1939, in Burke 2001:35-36). Historical relations between France and China, including the Sino-French war that was the backdrop for *Photo 1*, the potential material benefits of planes, and the symbolic value of hats and clothing are all part of the secondary level and the iconographic analysis of Panofsky. On the third level, that of symbolic value, we can mention the symbolic value of commerce and its link with friendship, world peace and the possibility of living in harmony with those who are very different from ourselves. In the words of Elizabeth Edwards, "photography is capable of providing that link between the lived experience, moving between personal and collective inscription, that is the raw material of history and memory and wider, more general structures" (2001:102).

6. The Interpreter Centre Stage

Photograph 7: The Interpreter Centre Stage

Photograph 7, taken at the Military Academy of Saumur on 23 May 1910, demonstrates the importance of Vissière, in the centre of the photo, again extremely elegant in top hat and tails, along with the Chinese delegation, alongside General Ha Han-Chang, while Prince Tsai Tao is with the French officers, thus demonstrating a certain integration. In "Interpreting Photographs" Elizabeth Edwards comments on the position of the interpreters at the edge in the photographs, "as signifying compositional elements which indicate the importance of the occasion and the political and photographic protagonists and as accoutrements to power and process" (2014:21), framing the photo and becoming "visible" in this setting only on the terms of the protagonists. Quite differently, here the diplomatic interpreter is right in the centre of the image. Again, the picture is one of harmony and a great improvement in Sino-French relations. Vissière is organizing the Chinese tour. He is the only one of the French group who knows Chinese. His diplomatic position gives him great authority. And diplomat-interpreters can have great prestige. In fact, he was called

the last representative of the line of the great "first interpreters", who, like the "first dragomans" of the embassies in Turkey, had not only the role of translating the dispatches but were also the real advisers ministers through their familiarity with a civilization very different from ours[10].

It seems that the position of diplomat-interpreter no longer exists. In Chapter 8 we shall meet Viktor Sukhodrev, a very successful interpreter for the heads of state of the Soviet Union, representing the interpreting profession in a more modern phase.

Currently, the main role of many diplomats is that of trade facilitator, and the commercial attaché is an important post in all diplomatic representations. Obviously, the Chinese military delegation was seen as a potential customer for French goods, here planes and airships. Aviation was becoming a highly profitable business, and competition was fierce. Vissière's interpretation was very successful as he managed to interest the Chinese delegation in the purchase of French aircraft. Shortly after this visit, and possibly to a great extent due to it, China would begin training pilots and creating its own air force. In October 1911 Zee Yee Lee was the first Chinese to obtain a permit to fly in the United Kingdom. However, they did not buy the planes from Blériot Aéronautique. In 1912, the Caudron brothers, who appear not to be present at this exhibition, sold twelve twin-engine aircraft to China and started the first flight school in China in 1913.

To summarize this chapter, it is important to mention that: (i) we have a vision of the interpreter as someone who unites nations and people; (ii) we get a glimpse of the excitement of the early days of aviation; (iii) the diplomatic interpreter has the role of facilitating the visit of the Chinese mission and introducing visitors to French aircraft manufacturers; (iv) Chinese visitors and aviation in China are of great interest to the French press, but interpretation and interpreters are not; (iv) the idealized "romantic" image of Vissière and Luo Fenglu can be contrasted with the documentary aviation photos from the press agency's newspapers; (v) the Issy photos are an excellent document of the apparently successful chuchotage work of a diplomat-interpreter; and (vi) these photographs depict very different but important historical moments.

---

[10] "[…] le dernier représentant de la lignée des grands "premiers interprètes", qui, tels les 'premiers drogmans (sic)' des ambassades en Turquie, n'avaient pas seulement pour rôle de traduire les dépêches, mais étaient les véritables conseillers des ministres par leur familiarité avec une civilisation très différente de la nôtre" (Pelliot 1930:27).

CHAPTER 2

THE SHOPKEEPER – JULIUS MEYER: "INDIAN INTERPRETER"

1. Introduction

Julius Meyer, born in Bromberg, Prussia, on 30 March 1839, was a Jewish immigrant who had arrived from Prussia in 1867 to join his older brother Max in Omaha, Nebraska, and in 1869 was followed by his brothers Moritz and Adolph. Max's general store prospered, and Max Meyer & Brothers Co. expanded, began selling jewellery and musical instruments, and for many years was the largest retailer of its kind in the western United States, earning it the nickname "Tiffany's of the West". It was the first Omaha store to have a telephone installed, a branch was opened in Cheyenne, Wyoming, and five travelling salesmen plied the territory west of Omaha to increase sales (Gendler 1968). In the 1880s, Max Meyer & Brothers Co. had 25 employees, with annual sales of $750,000 (in 2024 over $22 million), while annual sales for Moritz's businesses, which traded wholesale tobacco, cigars and sporting goods, totalled $350,000 (in 2024 some $11 million).

Nebraska prospered in the 1880s. While the state's population more than doubled in the decade between 1880 and 1890 Omaha's population increased from 30,000 to over 140,000, a 358% growth in ten years. Much of Omaha's growth was due to the development of meat packing plants (Gendler 1968). The creation of the Union Stock Yards Company in 1884 was a major factor in the development of the town's main industry and was accompanied by a growth in other businesses such as foundries, linseed oil, soap, bricks, garment factories, food processing, and distilleries.

Likewise, the number of Jewish immigrants in the USA increased: between 1820 and 1850, from about 5,000 to 50,000, and in the following decade the number reached more than 100,000. Political problems in Europe were the main cause of increased emigration of Jews to the United States.

Revolutions in Germany, Austria, and Italy resulted in large numbers of Jews disenchanted with life in Europe moving to America (Gendler 1968).

A small number of them went out west. In the words of Gerald Kolpan, who wrote a fictional biography of Julius Meyer:

> Almost from the beginning of Westward expansion, Jews made a home on the range. They were fur trappers, gold miners, cowboys, pedlars and scouts. There were sheriffs, marshals, mayors of small towns and at least one gunfighter. A *shana medele* [pretty girl] from San Francisco married Wyatt Earp; a storekeeper from Bavaria and a tailor from Latvia invented blue jeans (Kolpan 2012:Internet).

Upon arriving in the West, they managed to link these new economic markets to urban metropolises, introducing new actors, such as the Indians, and creating a new niche in the West for unknown Japanese and Chinese products, which were popular in the East, and, in return, also creating a new niche for indigenous products in the East, bringing exotic objects into urban homes and commodifying indigeneity (Koffman 2012). Native artefacts were shipped to museums, and buyers in small towns in the East could receive these indigenous products and artefacts by mail order (Koffman 2012).

The Jews acted as cultural mediators, translators, guardians, fighters, friends of the Indians and also historians, educators and ethnographers, as was the case of the German Jewish immigrant Franz Boas (Koffman 2012).

Once in Omaha, Julius quickly began to make contact with the Plains Indians. He managed to learn six Indian languages: Ponca, Omaha, Winnebago, Pawnee, Sioux, and Ogallala and apparently became good friends with the Indians, serving as an interpreter for General George Crook. In the late 1860s, he opened an Indian curio shop, called Indian Wigwam, on the corner of Eleventh and Farnam Streets in Omaha, selling beads, loafers, wampum bags, tomahawks, bows and arrows, petrifications, and peace pipes, among other products (Koffman 2012).

No longer working with his brothers, Julius began trading with indigenous tribes such as the Ponca, Omaha and Sioux. So well-known was he for his honesty that the Indians nicknamed him "Box-Ka-Re-Sha-Hash-Ta-Ka", "the curly-haired chief who speaks with one tongue" and made him their official interpreter. He was highly trusted by Chiefs Spotted Tail, Red Cloud, Sitting Bull and Swift Bear for his straightforward and honest way of doing business. At the same time, Spotted Tail, a Sioux, presented Meyer with his pipe and bag of tobacco, "the warmest declaration of friendship an Indian can make to a friend" (*Daily Herald*, Dec. 6, 1876, in Gendler 1968:19).

Living with Native Americans for weeks at a time, Julius was also famous for another quirk: following the rules of the Jewish diet. When he was invited to feasts with members of the tribe, his hosts knew how to serve him hard-boiled eggs instead of the non-kosher meat everyone liked, and he was once the special guest at a dog feast, "considered an honor for a white man" (*Daily Republican*, August 9 and 13, 1873; *Daily Herald*, July 24, 1877, in Gendler 1968).

According to Julius, in 1869 a hostile tribe attacked him. They tried to kill him – and it was only the intercession of Standing Bear, chief of the Ponca, that saved his life. Julius became the interpreter for Standing Bear, and was soon translating for famous chiefs such as Sitting Bull, Red Cloud, and Swift Bear (Kolpan 2012).

Meyer closed his "Indian Wigwam" in 1880 but maintained his association with the Indians. In 1883 Julius Meyer was commissioned by the French government to accompany a group from Omaha and Winnebago to Paris (Gendler 1968). For many years Meyer served as an agent for the Omaha government's indigenous groups, often fighting for native rights. Another scheme was to take the Standing Bear and a group from Ponca on a year-long tour of the Paris Exposition of 1889, where they were quite successful. He later had a short career as an insurance sales representative (Gendler 1968).

Julius Meyer was a founder of the Omaha Musical Union and a member of Concordia, a German singing society. He was involved in the building of the first synagogue in Nebraska, the Omaha Congregation of Israel (now Temple Israel), and the Omaha Hebrew Benevolent Society. In 1886 he organized a Young Men's Hebrew Association in Omaha. This association sponsored social and literary entertainments. And he was also part of the Metropolitan Club, formerly the Centennial, and later the Standard Club (Gendler 1968).

Max Meyer & Brothers Co. went bankrupt because of the Depression of 1893, and Max and Adolph Meyer left Omaha. Max got a job selling jewellery in Baltimore (Gendler 1968). Julius Meyer, who never married, was found shot dead in Hanscom Park, Omaha, in 1909 under highly mysterious circumstances. Apparently a suicide, it was reported at the time that he killed himself, first by shooting himself in the temple and again in the chest, with his left hand, although Julius was right-handed. No alternative theories were advanced, and his death was mourned throughout the American West as a tragic case of suicide (Gendler 1968).

2. Welcome to Capitalism!

Photograph 1: Welcome to Capitalism (c.1873)

This photograph, probably from 1873, shows Julius Meyer in his first store, Indian Wigwam, at 163 Farnam St., Omaha. Julius, in European attire, dominates the photo. He looks in good spirits and very much at ease, while the Indians, from the Pawnee group, with Chief Petalesharo II on Julius's right (Amertribes), all look very miserable and seem to have agreed to be photographed with some reluctance.

The image gives us an idea of some of the goods Julius traded: buffalo hides, cradleboards used by indigenous women to carry babies, machetes and the peace pipes held by the Indians.

Photograph 1a: Welcome to Capitalism, with framing (c.1873)

Formally, the image is a series of concentric frames (*Photograph 1a*): the two titles, "Indian Wigwam" and "Julius Meyer 163" announce the subject of the photo. The shop front frames the buffalo hides and cradles. These, in turn, frame the door, which frames the stage, with Julius in the centre, flanked by the Indians. The participants form a triangle, with Julius at the apex, in the most important position (Kress & van Leeuwen 2006)

We can see this photo as a *studium* of harmonious relations between merchants and Indians, demonstrating that Indians were successfully introduced into North American capitalism. But the *punctum* destroys this harmony: why do the indians look so morose? Unlike the images of the Chinese military mission that we saw in Chapter 1, there are no smiles here. Were the Indians photographed willingly? Were they paid? Was this photo taken as a condition of a business agreement? Or does this ill will come from a cultural reluctance to be photographed? Julius' hand is on the shoulder of one of the Indians. Is he being condescending? Or is this a sign of support and friendship? Or is he holding the Indian so he doesn't move and ruin the picture? What is the relationship between Julius and the Indians? Was it really as friendly as I reported above, using information from Carol Gendler's dissertation, "The Jews of Omaha: The First Sixty Years" (1968), based on contemporary Omaha newspapers?

We can examine this point on several levels. Such gloominess can be interpreted differently. For Melissa Block, images such as Edward Sheriff Curtis' photographs of Native Americans from the early 1900s are "very dignified, very prideful, unsmiling, yes, but most photographs from the early 1900s, people would not be smiling. It was a serious business to have your picture taken" (The Picture Show: Photo Stories from MPR 2011: Internet)[11].

Standing still, in many cases for a few minutes, in the early days of photography was not easy, and therefore, when Julius appears holding an Indian, he was probably trying to stabilize him so as not to blur the image. Smiling in a photo is relatively recent. And keeping a smile for several minutes is far from easy.

Going into a studio to have a photo taken was often considered a daunting task. For Wilkie Collins (1824-1889), life's three great trials were "Having a tooth out, having your hair cut, and having your photograph taken", and Thomas Carlyle (1795-1881) described the preparations necessary for a portrait by Julia Margaret Cameron (1815-1879) as "a kind of inferno" (in Pols 2015).

Furthermore, various cultures, including Native American cultures, believed that the camera could steal people's souls. The native warrior of the Oglala Lakota group, Crazy Horse, would not allow anyone to take his picture, not even at his own funeral! (Mansbridge 2016)

In "Navajo and Photography" James Faris examines the relationship of the North American Navajo peoples in the US Southwest and photography, emphasizing that they "frequently consider untoward photographs (those that they did not agree to have made, or over whose distribution they have had no control) to constitute a danger to them by their very existence" (2003:89), and the Navajo have filed a number of lawsuits in order to prevent the publication of photographs they did not agree to have taken.

---

[11] Melissa Block was interviewing Ryan Red Corn, an Osage Indian from Oklahoma, who had just shot a four-minute film on "Smiling Indians".

The great majority of the photographs taken of the Navajo and other native peoples, as we can see in this chapter, are produced by non-indigenous peoples and based on the traditional clichés and antitheses of "civilized/savage, sedentary/nomadic, modern/traditional, anthropological knowledge/local ideology, and West/Other" (Faris 2003:89). Faris tries to overturn this bias by presenting artwork and photographs made by Navajo. In addition, there is an attractive photo of a Navajo woman with a broad smile taken by Curtis in 1907. However, this was never published though a photograph of the same woman with a dour expression was (Faris 2003:94).

We can return to Rachael Langford's article "Framing and Masking", in *Framing the Interpreter*, which I mentioned in Chapter 1. In Figure 3.3, she contrasts the activity of the interpreter with the passivity of native women in the Dakar postcard photograph taken between 1902 and 1904:

> […] the women and children form a tableau of passive reception – they are not strictly engaged in any particular activity – while the interpreter extends tobacco leaves to them. Thus the image depicts the indigenous women as fixed, "being", against the interpreter's image of "doing" (Langford 2014).

Here we find much the same thing: Julius is the European colonizer, the successful shop owner, the man of action, buying and selling, making money from the Indians but also helping them to enter the capitalist world; however, they seem to have little enthusiasm for this brave new world and are silent and sullen.

In *Photography's Other Histories* (Pinney and Peterson 2003), which contains the above-mentioned article, "Navajo and Photography" (Faris 2003), two chapters describe and illustrate photos taken by groups traditionally considered "subaltern". In "Growing up with Aborigines" Michael Aird describes his 1991 exhibition "Portraits of Our Elders", showing 1920s photos going against the grain of depicting aboriginal poverty and which, taken inside studios, showed "well-dressed and confident Aboriginal people who sent their children to schools to be educated in order to improve their credibility or acceptability within the European society" (Aird 2003:25). And in "Yoruba Photography: How the Yoruba See Themselves" Stephen F. Sprague describes the ways in which photography has been a part of Yoruba culture since the 1930s (Sprague 2003).

The concept of photography as liberating rather than oppressive can be seen in Frederick Douglass (1818-1894), born into slavery and who became an abolitionist, suffragist, public speaker, writer, government official, and civil rights activist and leader. He believed in the role of photography as an aid to achieving citizenship for black people and amassed more than 160 personal photographs and was even called the most photographed man in America in the 19th century (Cain 2017). Photographs as a form of representation insisted on the fact that those portrayed had a role to play in society. They were a democratic social leveller, the aftermath of slavery: "The humbled servant girl whose income is but a few shillings per week may now possess a more perfect likeness of herself than noble ladies and court royalty" (Stauffer et al. 2015; Hirsch 2020:112); and: "Men of all conditions may see themselves as others see them" (Stauffer et al. 2015). Indeed, by the mid-19th century, photographic studios could be found even in small US towns. Douglass' own stern and imposing poses in front of the camera were also a way of demonstrating the dignity of black people, for Laura Wexler, "a 'revenant' from the social death of slavery" (Wexler 2012, in Hirsch 2020:113). "For Douglass, photography was the lifeblood of being able to be seen and not caricatured, to be represented and not grotesque, to be seen as fully human and not as an object or chattel to be bought and sold" (Bernier 2015 in Stauffer et al. 2015, and in Cain 2017).

3. Julius Meyer – "Indian Interpreter"

With the huge letters on the store's awning, "Julius Meyer – Indian Interpreter", Julius becomes a very visible interpreter in this photo, taken in 1878, and in *Photo 3* we see that his new premises occupy two floors and are, in fact, quite spacious and in a more solid brick building, with the word "CURIOSITIES" painted in large letters on the side wall, preceded by "Indian, Chinese & Japanese", and underneath the last "S" of "Curiosities", "Cigars, Tobacco, Cutlery, Notions". It looks like the store of a well-off and successful merchant who took advantage of the growth and new wealth of the town of Omaha to market non-essential goods, oriental and native curiosities, to a population with money to spend. The large "Indian Interpreter" on the awning may have been a marketing device, communicating the personal contact he had with indigenous groups, and it appears that he was also known to many as the "Indian Interpreter", which was something of a nickname. The crowd in front of the store appears to be a mixture of Indians and whites, with Julius at the front in a dark shirt. The impression is that anyone could join the crowd and be in the picture, and this harmonious crowd scene shows us the popularity of the Indian Wigwam and the mix of races and social classes.

242½ CIGARS & TOBACCO. 24
JULIUS MEYER.
INDIAN INTERPRETER
IMPORTED KEY WEST & DOMESTIC CIGARS.

Photographs 2 & 3: Julius Meyer – "Indian Interpreter"

4. Julius Meyer and the Pawnee Chiefs, c. 1869

Photograph 4: Julius Meyer and the Pawnee Chiefs, c. 1869

The photo vectors again form a triangle, with Julius once again dominant and in the centre, at the apex of the triangle. He is handing a peace pipe to the man on his right. This time they seem to be interested in what he has to say, and his role is that of teacher and authority.

The photograph was taken in Frank F. Currier's studio in Omaha, and Meyer is wearing a buckskin scout's coat with collar and cuffs of otter fur and a turban made of striped skunk fur. These clothes could have been Julius's or provided by the studio as it was common to "dress up" to take pictures. For this photo, in which the group are formally posing, Julius assumes a hierarchical position of omniscience, of the teacher, the authority, the representative of the superior white culture that conquered the indigenous peoples and made them admire him. He is at a higher level than the Pawnees, and we can return to the spatial distribution: in the upper part of the photo, as we saw in Chapter 1, there is an attempt to escape gravity, terrestrial life, to rise to the sky and clouds (Arnheim 1974). The condescending manner in which Julius hands the Pawnee the peace pipe contrasts with the harmony and equality seen in *Photo 1* of Chapter 1, with Vissière and Luo Fenglu smoking together. The peace pipe he is handing to the chief on the right has great semiotic value, and the message could be: "We of the white race are superior, but we come in peace and we want you to learn from us and participate in our (capitalist) world to improve your lot". Two Pawnee boys, perhaps the representatives of the new mixed and integrated American society, sit in the foreground.

5. Carte de visite and Julius Captured and Scalped!

Photograph 5 is another studio portrait, in a pseudo-natural scene, with Julius again wearing a raccoon fur hat, typical of frontiersmen. He's touching Spotted Tail, perhaps as a gesture of friendship, but probably to try to keep him from moving and spoiling the shot, which would take several seconds. He appears to be much more respectful of the Sioux-Lakota than he is of the Pawnee and Ponca in the previous photos, and we must remember he is facing the chief with the fateful name of Pawnee Killer! This double photograph was supposed to be viewed with a stereoscope, which gave the image a three-dimensional effect.

Photographs of staged theatrical scenes, the most famous of which were Julia Cameron's 1874 photographs accompanying Alfred Lord Tennyson's poem *Idylls of the King*, were popular in the 19th century, and photographic studios would have a selection of costumes for such scenarios. Here, in *Photograph 6: Julius Captured and Scalped!* we see Julius enacting a scene in which he himself is scalped by the Pawnee Indians. However, only Julius seems to be getting into the spirit of it as the Indians look on with little enthusiasm.

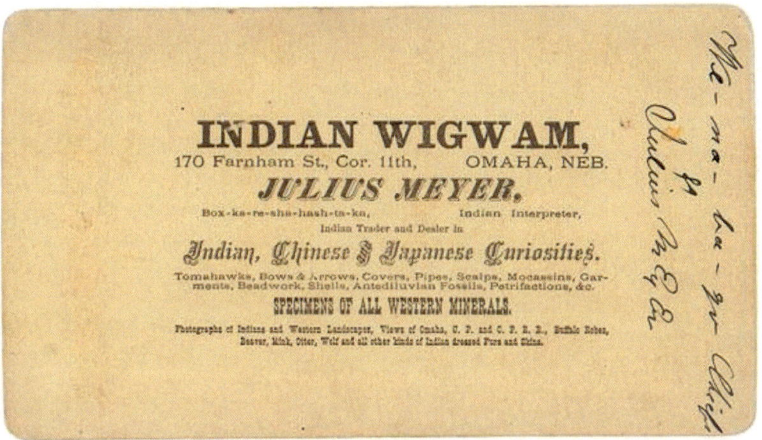

INDIAN WIGWAM,
170 Farnham St., Cor. 11th,   OMAHA, NEB.
JULIUS MEYER,
Box-ka-re-sha-hash-ta-ka,   Indian Interpreter,
Indian Trader and Dealer in
Indian, Chinese & Japanese Curiosities.
Tomahawks, Bows & Arrows, Covers, Pipes, Scalps, Moccasins, Garments, Beadwork, Shells, Antediluvian Fossils, Petrifactions, &c.
SPECIMENS OF ALL WESTERN MINERALS.
Photographs of Indians and Western Landscapes, Views of Omaha, U. P. and C. P. R. R., Buffalo Robes, Beaver, Mink, Otter, Wolf and all other kinds of Indian dressed Furs and Skins.

Photograph 5: Carte de visite. With Lakota Chiefs: Spotted Tail (Brulé-Lakota-Sioux), Iron Bull (Crow-Wyoming) and Pawnee Killer (Ogala Sioux), c. 1870s

Photograph 6: Julius captured and scalped!

We may here draw a parallel with the salvage and reenactment of indigenous ceremonies photographed and filmed for the benefit of western anthropologists, which is a central theme of Elizabeth Edwards' *Raw Photography* (2001), in which she discusses the reenactments in the 1898 expedition to Torres Strait Islands, off the northernmost point of Australia. Such reenactments were usually made due to the limitations of early photographic and filming techniques and the desire to produce reliable academic material of the rituals of the peoples, especially the death of the traditional Torres Island hero Kwoiam. She also mentions the reconstructions of Franz Boas (1858-1942) for the camera of Kwakiutl technologies and cultural activities in British Colombia published in 1921 (2001:170) and the great photographic of work of Edward Sheriff Curtis (1868-1952) on Native American cultures between 1900 and 1930 (2001:171), also based on reconstructions.

But is this parallel really valid? Julius is a Jewish immigrant, hardly a member of the dominant American WASP class. Although he seems to have some control over them, his Pawnee participants are obviously reluctant actors, and the scene degenerates into a comic farce and sham!

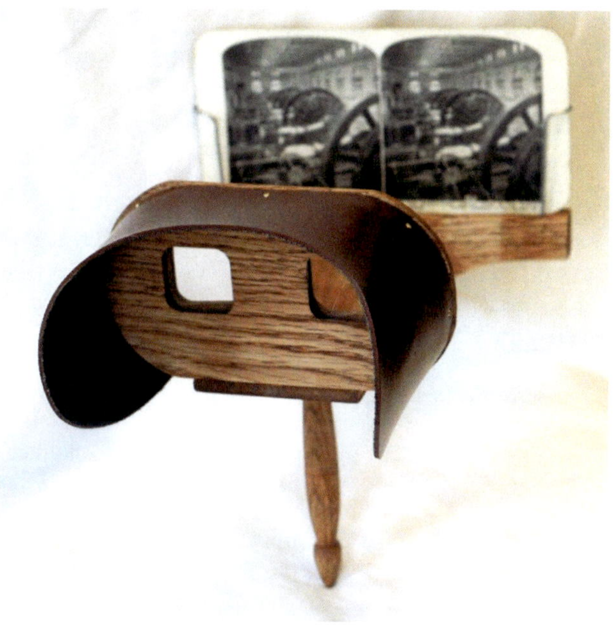

Figure 1: A Holmes stereoscope, the most popular brand in the 19th century.

Relations between the Sioux or Lakota and the Pawnees were poor. The Battle of Massacre Canyon took place on 5 August 1873 in Hitchcock County, Nebraska. It was one of the last battles between the Pawnee and the Lakota Sioux, and the last major battle between groups of North American Great Plains Indians. The Pawnee had been under attack by the better-armed Lakota for many years, and from the 1840s to the 1870s they were severely weakened by famine, disease, and warfare with other Plains Indian tribes. In order to survive many Pawnee warriors joined the United States Army as scouts and took part in the Army's campaigns against the Sioux. The battle occurred when a group of Oglala/Brulé Sioux with over 1000 warriors attacked a group of Pawnees who were hunting summer buffalo. Over 70 Pawnee warriors, plus over a hundred women and children, were killed. Many of the victims were brutally mutilated or scalped. As a result the Pawnees decided to cede their land in Nebraska to the government and move to a reservation in Oklahoma (Correa 2013). Julius appears to be on good terms with both groups, but in neither photo do we see members of the two groups together, and his relationship with each appears to be very different. The enmity can clearly be seen in the fact a Lakota was actually called Pawnee Killer.

The photograph was used as a carte de visite (CDV), with details of the Indian Wigwam Store on the back. Cartes de visite were very popular at the time, and many must have been distributed by Julius in order to attract business. This photograph is mounted on an imperial-sized CDV, measuring approximately 9.5 x 6 cm, on top of rigid piece of cardboard measuring approximately 10 x 6.5 cm. The names of in the photograph are identified in purple ink: "Julius Meyer Interpreter, & Indian Chiefs. Spotted Tail, Iron Bull, and Pawnee Killer". And on the back: "INDIAN WIGWAM / 170 Farnham St., Cor. 11th, OMAHA, NEB / *JULIUS MEYER* / Box-ka-re-sha-hash-ta-ka. Indian Interpreter. / Indian trader and dealer in / *Indian, Chinese & Japanese Curiosities* / Tomahawks, Bows & Arrows, Covers, Pipes, Scalps (sic!) Mocassins, Garments, Beadworks, Shells, Antediluvian Fossils, Petrifications &c./ SPECIMENS OF ALL WESTERN MINERALS / Photographs of Indian and Western Landscapes, Views of Omaha,U. P and C. P. E. R., Buffalo Robes, Beaver, Mink, Otter, Wolf and all other kinds of Indian dressed Furs and Skins".

The exchange of cartes de visite was a great fad in the middle of the 19th century and gave rise to another fashion: photograph albums. Cards were exchanged between friends, family, and collectors. "Visiting cards, as everyone knows, have become the social currency, the dollars of civilization". This sentence by the American writer Oliver Wendell Holmes (1809-1894) was written in 1863 and demonstrates the popularity of this photography format around the world (Wanderley 2016).

The camera for taking pictures on business cards was patented by André Adolphe-Eugène Disdéri (1819-1889) on 27 November 1854, under the name of carte de visite: a camera with four lenses to obtain eight portraits on just one glass plate; the first four photos were exposed, the plate moved and allowed the other four photos to be exposed. The copying was generally done using the albumin printing technique. The invention allowed, for the first time, the mass production of photographs.

The height of popularity of cartes de visite began in 1859, when, on the same day that he was leaving for Italy, 10 May, Napoleon III stopped at the well-known studio of Disdéri in Paris, to have his portrait taken. The next day there were already queues outside the studio. The enthusiasm surrounding these portraits was enormous, and Disdéri's luxurious studio, befitting his reputation in the 1860s as the richest photographer in the world, was described by a German traveller as the true temple of photography. Also, according to this report, Disdéri had a daily turnover of 3,000 to 4,000 (about $35,000 to $50,000 in 2024) from the sale of portraits and employed

90 people who took more than two thousand photographs a day (Wanderley 2016).

Figure 2: Multi-lens camera introduced by Disdéri for the production of photographs in the carte de visita format.

In New York, the cartes de visite probably began to circulate in the summer of 1859, introduced by the photographer Charles DeForest Fredricks (1823-1894). The North American Civil War (1861-1865) helped popularize them as soldiers and their families had their portraits taken before being separated by the conflict. In Great Britain Queen Victoria (1819-1901) completed more than 100 portrait albums of members of the royal family and important people from society.

The decline of the carte de visite format took place from the 1870s onwards, when it began to be supplanted by the cabinet card, which appeared in England in 1866, which was slightly larger, with photographs of around 9.5 x 14 cm on rigid pieces of card of about 11 x 16.5 cm (Brasiliana Fotoográfica).

6. The Washington Delegation of 1875

Photograph 7: Washington Delegation 1875

This photograph of the 1875 Indian delegation in Washington, DC was taken by Frank F. Currier on 13 May 1875 in his studio in Omaha, Nebraska. The delegation was about to leave for Washington to meet with President Grant to discuss the Black Hills area.

Julius is next to Red Cloud, the chief of the Ogala, and in front, from left to right, we have Sitting Bull, another leader of the Ogala, not the more famous Hunkpapa Sitting Bull, as well as Swift Bear and Spotted Tail, both Brulé chiefs. All belong to the Lakota ethnic group, the Western Sioux. In *Red Cloud: Photographs of a Lakota Chief*, Frank Henry Goodyear writes that this was Red Cloud's third trip to Washington and he intended to discuss three things: he was unhappy with the behaviour of John J. Saville, the new head of the Red Cloud Agency; he was also angry about the recent expedition of gold prospectors led by George A. Custer; and he also wanted to discuss the government's proposal to move the Lakota settlement to a federal Indian reservation. For its part, the government wanted to discuss the possible purchase of the mineral-rich Black Hills from the Lakota (Goodyear 1967).

On the surface, this photograph a harmonious *studium* of a group together with the interpreter, but the *punctum* is: why is Red Cloud wearing western clothes? Goodyear notes that a local clothing store gave Red Cloud a formal suit, which he is wearing here, and which, according to the *Omaha Republican*, aroused the jealousy of the other chiefs. Meyer was the host of the group in Omaha and covered their expenses. However, they demanded payment of two ponies for each of them for having their pictures taken (Goodyear 1967).

Furthermore, Goodyear tells us that the feud between Red Cloud and Spotted Tail, with both disputing the position of the main spokesperson for the Lakota, was already over although Red Cloud's new suit seems to have given him precedence.

The photograph is in the cabinet card format, which usually measured 10.8 x 16.5 cm, and the printed inscription gives it a certain authority. At that time, in all urban centres there were already a large number of photographic studios, with very different prices, from the simplest to the most sophisticated studios. Currier's studio, as it was chosen to take the official photo of the group going to Washington, was certainly the most prestigious in Omaha.

7. More Indian Interpreters from the 1875 Washington Delegation

Photograph 8: More Native Interpreters from the 1875 Washington Delegation

This photo was also taken by Frank F. Currier on 13 May 1875 in his studio in Omaha, Nebraska. Julius has now changed clothes and is wearing the formal three-piece suit he would most likely wear when visiting the President in Washington. With him we see two other indigenous interpreters, and there is a certain bond between the three. Standing, left to right: Louis Bordeaux (1849-1917) (interpreter at the Spotted Tails Agency), son of merchant James Bordeaux and Huntakalutawin (Marie Bordeaux) (1822-1884), sister of Chief Swift Bear of the Brulé Lakota; William Garnett (Red Cloud agency interpreter) (1855-1930), son of Major Richard Brooke Garnett and Ogala Looks-at-Him (Mollie Campbell); Julius Meyer (Interpreter and Local Host in Omaha). Seated from left to right: Sitting Bull (Oglala-Sioux), Swift Bear (Brulé-Sicangu-Sicangu), Spotted Tail (Brulé-Lakota-Sioux-Sicangu), Red Cloud (Oglala-Lakota-Sicangu).

While *Photo 7* appears to have been used commercially by Julius as a cabinet card and distributed to clients, acquaintances, and business contacts, *Photo 8*, according to Frank Goodyear, was the official photo of the group traveling to Washington. With Julius now wearing his suit like the other interpreters and Red Cloud no longer wearing his western suit, the two groups are clearly separated by their clothing.

Goodyear notes that the government agents accompanying the trip did not appear in the photo, pointing to their lack of political importance, although visits to photographic studios – like the trips they were part of – were in fact highly significant diplomatic manoeuvres: "agents lavished attention on the native leaders, all the while procuring images that could serve a variety of purposes", and indigenous leaders entered into such exchanges, believing that this could help them (Goodyear 1967:30). However, little was accomplished in Washington: President Grant and the head of the Commission on Indian Affairs, Edward P. Smith, were unmoved by Red Cloud's complaints, and the Lakota leaders were offended by the government's offer to buy the Black Hills. On the last day, Red Cloud refused to have his photograph taken by the New York Graphic, demanding US$25 [in 2024 rather more than US$500!]

Here we see a very different Julius. In this photo, probably composed with great care by Frank Currier, who must have been delighted to receive this commission from the US government, Julius takes his place along with the other interpreters in a position of equality and friendship, with their hands on each other's shoulders, though this could also have been a way making sure they all kept still. Julius stands out as non-indigenous as the physical traits of Louis Bordeaux and William Garnett clearly show their indigenous origins. In practice, it seems more convenient and polite to allow the older Indians to sit, and there was less chance of them moving and spoiling the photo, while the younger and fitter interpreters remained standing, but we can also interpret the photo using the ideas of Arnheim (1974) and Kress and van Leeuwen (2006) on the top and bottom of an image. The lower part of the image has greater weight and represents the "real", the "present", and the upper part is the "possible", the "future", the "imagined". Thus, the future lies with the mixed young population the ethnically mixed Bordeaux and Garnett, and the immigrant Meyer who, with their American and European values, seen here in their Western attire, will leave their indigenous culture behind for a better future.

8. Final Words

Very different to the photographs of Arnold Vissière, which showed us a certain equality in the relations between France and China, the fascination with China in *Photo 1*, the trade between France and China in the series of photos related to aviation, and a harmonious "contact zone", Julius Meyer's photos are closer to Mary Louise Pratt's description of the contact zone as:

> "[…] the space of imperial encounters, the space in which peoples geographically and historically separated come into contact with each other and establish ongoing relations, usually involving conditions of coercion, radical inequality and intractable conflict".

Perhaps Julius's dealings with the Pawnee and Lakota Indians reflected the "soft power" of military force and American capitalism as the Indians are coerced into commerce whether they like it or not. Much of their native land is occupied, and so they are compelled to depend on intermediaries like Julius, who will bring them a few dollars for curiosities sent to museums and to decorate middle-class homes in the East.

Julius's photographs have an obvious theatricality, and we can associate them with the North American sociologist Erving Goffman's ideas on the theatricality of everyday life (Goffman 1990). We play various roles in everyday life, manipulating other people and changing our behaviour to suit our interlocutors. With the Pawnee Indians, Julius plays the role of a mentor, king, and god. He manipulates them into acting, pretending to be captured and scalped, although the Indians seem not to be amused. As mentioned, many Pawnees were killed by the Lakota at the Battle of Massacre Canyon in August 1873, and their status appears to be much lower than that of the Lakota groups, but the Pawnees still retain their pride. However, Julius's attitude towards the Lakota Indians is much more respectful, and he approaches them from a position of equality, not superiority, as in the Pawnee pictures. Another photograph, which is not here, shows Julius in a tailcoat, three feet away from Standing Bear, who is wearing a headdress, the two of them standing and looking at each other. Goffman also sees life as a stage (1961): depending on the situation, we can be in the centre of the stage, in the foreground or background, on the sidelines or even behind the scenes. In the pictures with the Pawnees, Julius is clearly centre stage; but he's just a supporting actor in the Lakota pictures.

We dress according to the role we want to play. Julius changes his role and costume depending on the people he is dealing with, and Red Cloud puts on and takes off his new suit. In Chapter 1, at the Issy-les-Moulineaux airfield, the participants dressed according to their role, and Vissière's top hat and tailcoat asserted his authority.

We saw Julius Meyer as a mediator between the capitalist system, the US government, and various groups of Indians in Omaha. We've seen Julius in various poses and costumes in photographs taken with Indians. We've also discover the importance of cartes de visite, widely distributed by commercial establishments and individuals and, from the 1880s, the importance of cabinet cards, which, in the 1900s, were supplanted by postcards.

Like Chapter 1, this section emphasized the positive role played by interpreters. However, in this case, the frowning faces of the indigenous subjects may undermine this proposition.

Chapter 3

The Interpreters of the SPI (Indian Protection Service)

1. The Photographic Archive of the SPI (Indian Protection Service)

The work currently carried out by the Fundação Nacional dos Povos Indígenas [National Foundation of Indigenous Peoples] (FUNAI) was previously performed by the Serviço de Proteção aos Índios [Indian Protection Service] (SPI), the official body responsible for coordinating and executing the federal indigenist policies of the Brazilian government and protecting and promoting the rights of indigenous peoples in Brazil.

The SPI was created in 1910 to intervene in the conflicts during the colonization of the west of the state of São Paulo, which were decimating the indigenous peoples of the region. Marechal Cândido Rondon was invited to organize the SPI due to his experience in expeditions in the Brazilian interior acquired during the activities of the Rondon Commission, a 15-year project that aimed to link the federal capital at the time, Rio de Janeiro, to the remote areas in north and western Brazil through telegraph lines. Instead of the extinction and genocide of indigenous peoples, Rondon advocated a policy of pacification reinforced by the motto "Die if necessary; but never kill"[12] (Freire 2010:11).

The vast majority of official SPI photos, especially those published in the book *Memória do SPI: textos, imagens e documentos sobre o Serviço de Proteção aos Índios (1910-1967)* [Memory of the SPI: texts, images and documents on the Indian Protection Service (1910-1967)], reflect Rondon's policy of integrating indigenous groups into Brazilian society as rural agricultural workers and to eventually lose their indigeneity, seen as inevitable by Rondon's positivist ideas: they would learn the colonizer's

---

[12] "Morrer se necessário; matar, nunca"

way of life, customs, religion and language, and they would receive the advantages of "civilization" and integrate into Brazilian society.

This would be the solution to prevent the genocide of many indigenous groups, including the Xetá, as we will see in this chapter. Thus, the photos we see in the *Memória do SPI* are of school groups, Indians working as agricultural labourers, girls in sewing groups, boys playing football, all well-behaved and dressed and ready to become integrated into the lowest rungs of Brazilian society. They are essentially colonial photos, a "showcase" of Brazilian state intervention in the situation of indigenous groups, as pointed out by Freire (2011:17).

In fact, one has to dig deep into this carefully produced and illustrated book to find an image of naked, sick, or malnourished indigenous men and women. These few photos are of the misfits, the laggards, the groups that have lost the race for now but will soon take their rightful place at the hoe, in the kitchen, or at the sewing machine. The SPI images focus on the progressive project of the Brazilian state and hides the existing reality of suffering and disputes.

However, in the larger collection of the SPI archive, we can find a series of pictures of interpreters, some of which challenge the official position of the SPI and, looking deeper into others, finding the *punctum*, Barthes' term, which we have already seen in previous chapters, the disconcerting element that can appear in a *studium*, a study of a theme (Barthes 2015), here the success of the SPI in integrating Brazilian Indians into Brazilian society. In this chapter, we study several of these photographs, which show the contrasting elements of the SPI and its interpreters.

2.1. The Expeditions to Find the Xetá

In October 1955, the SPI formed an expedition to search for members of the Xetá group, some of whom had turned up at Fazenda Santa Rosa. In this photo the expedition is stopping for a welcome rest and cigarette break. The expedition members are looking directly at the camera, at least the main participants, the three men in the centre of the photo: with a handkerchief on his head, Antônio Lustosa de Freitas, owner of Fazenda Santa Rosa, which was part of a larger area of owned by his uncle, state deputy Antônio Lustosa de Oliveira; in the middle, José Loureiro Fernandes (1903-1977), Professor of Anthropology at the University of Paraná, coordinator of the research carried out by the expedition; and on our left, Dival José de Souza Nenê (1925-2005), an employee of the SPI and organizer of the expedition's logistics (Silva 2005:199). Although other photos show that Loureiro is much shorter than Lustosa and Dival, here, sitting on the ground, they look quite similar: Latin features of Portuguese or Spanish origin: dark hair with moustaches, burly. They're sweating, they look totally out of place in the forest, and they are suffering.

Photograph 1: Blokes Suffering in the Forest (1955)

There is a central core in the photo consisting of the three expedition organizers. The sweaty men form an inner circle, outside which are the two indigenous Xetá boys, Tuca, nine, and Kaiuá, younger, forming a group to our left, appearing quite distant from the others, and the other scattered individuals: Loyola Neto, journalist, smoking in the background; the photographer Ferralma probably left the camera mounted on the tripod to take the photo with the self-timer; and the only figure to appear calm, without heat and at ease, Loureiro Fernandes' student, Ney Barreto. Although the boys were taken away from their families and now live in Dival's father's house, they seem to have very little contact with any of the white men, and there certainly doesn't seem to be a surrogate father-son relationship with any of them. In addition, nobody thought of putting the boys in a central position in the picture.

Very different from Vissière's photographs in Chapter 1, in which we see great contact between the participants, here there are no vectors between the participants. Everyone seems isolated and closed within their own worlds. If we enter the photo on the left, we see the two boys: Tuca looking at the camera, but with a frown, visually separated from the group by the tree trunk, and Kaiuá is more interested in something he is holding. Here we can even make another division: on one side, the two indigenous Xetá boys, and on the other, the white men, with a clear division between the two groups, and everyone is cloistered in their own worlds, resting, smoking, or looking away.

The photo has something of a holiday outing, or perhaps an excursion gone wrong. Lustosa's knotted handkerchief on his head reminds me of my childhood in England in the 1960s. Few British men, at least in the working class, would have or need a sun hat, and the solution to the occasional hot day on the annual outing to Blackpool, Weston-super-Mare, or Skegness, was to protect the scalp with your cloth handkerchief, very similar to Lustosa's, normally used to blow your nose.

2.2. The Xetá Indigenous Group

The Xetá were the last ethnic group in the southern Brazilian state of Paraná to have contact with Brazilian society. The first recorded contact took place on 6 December 1954, when six men, tired of constantly fleeing the colonization fronts that had advanced in their territory since the end of the 1940s, drastically reducing the group, visited the Santa Rosa farm of Antônio Lustosa de Freitas. These men then withdrew into the forest, returning a few days later with their families, including women and children. Then visits became frequent.

In 1955, upon learning of direct contact between the Xetá and the inhabitants of Fazenda Santa Rosa, the SPI organized three contact expeditions in the region, known as Serra dos Dourados, one in October and another in November 1955, and a third in February 1956. *Photographs 1* and *2* are from the first expedition. The second and third expeditions managed to find some Xetá Indians.

In February 1956, an expedition from the Federal University of Paraná, led by José Loureiro Fernandes, found two other villages in the forest. No other members of the first expedition to search for the Xetá participated in this search; nor were they sighted by subsequent expeditions. No other groups were subsequently found. However, the family that made contact with the people at Fazenda Santa Rosa continued to visit.

2.3. The Interpreters Tuca and Kaiuá

Tuca, or Tucanambá José Paraná, was captured in 1953 at the age of six by surveyors from Colonizadora Suemitsu Miyamura & Cia. Ltda. and SPI representatives when he was picking fruit from a tree[13] (Silva 2005:198). He was taken to the home of Deocleciano de Souza Nenê, known as Nenê, an SPI inspector, who took care of Tuca in his guest house in Curitiba. There he met Kaiuá, Antônio Guairá Paraná, who had been captured in 1952 (Silva 1998:45). Tuca started studying but never finished primary school, only studying up to the second year. Little by little he began to learn about "white" life and to acquire the necessary skills to work in the guest house. He also took care of Kaiuá, who never managed to adapt to life outside the forest, barely learned Portuguese, and didn't even know how to wash his clothes. Tuca became a father figure for the other Xetá children who lived with the whites, the boys Kuein and Kaiuá and the girl Ã. He was baptized, and his godparents were the governor of the state of Paraná, Moysés Lupion and his wife. According to Dival, this was apparently a tactic by Lustosa and Nenê to gain Lupion's support for the cause of the Xetá people, but it was unsuccessful (Silva 1998:47). Then, from October 1955, after three years living with the whites, Tuca was invited to accompany several research expeditions, together with Kaiuá, and he did so until February 1961.

With the closing of the guest house in Curitiba, he went with Dival to the Guarupuava Indigenous Post, where he became Dival's cook. There, he met up again with his sister A'ruay, his brother-in-law Eirak, and his nephew Tiquein, in addition to Kuein and Kaiuá. Tuca went to the Manguerinha Indian Post, where he met his first wife, worked at a sawmill, returned to the Guarupuava Indigenous Post, met and separated from his second wife, a Kaingang woman, and finally met his third wife, with whom he lived at the Rio das Cobras Station, where he worked (Silva 1998:52).

He died aged 61 on 11 June 2007 (Peres 2007), one of the last speakers of the Xetá language. In 1999 the number of Xetá speakers dropped to eight although there were more descendants of Xetá who were unable to speak the language (Povos Indígenas do Brasil, 2019).

---

[13] The idea of "capturing" indigenous children is strange and repugnant to us today. It is understood that they were taken to "save" them, and they would act as a bridge between the SPI and the Xetá.

Photograph 2: The Interpreter at Work (1955)

This photo was also taken during the first Xetá expedition. Here we can see the two young boys helping Loureiro Fernandes to identify an artefact found in the Xetá village. Unfortunately, the type of artefact is not clear from the photo; it may be one of the stone tools used by the group and which are now part of the Loureiro Fernandes collection at the University of Paraná, Brazil (Merencio 2015).

However, the focus of the image is not on the object, but rather the interaction between Loureiro Fernandes and the boys, especially Tuca, who is always the one leading the interpretation. Tuca's job was to interpret for all the Xetá individuals the expedition encountered, also to work as a guide, and when they found the abandoned villages, to provide details about everything they found. According to official reports, he took a proactive stance. In this photo, for example, Tuca is doing his best and seems to remember a lot of things. Once more Kaiuá is at a distance, again dependent on Tuca.

Therefore, Loureiro depends on Tuca as one of the few surviving speakers of Xetá, a very difficult task for a nine-year-old boy. Though he may not realize it, he is responsible for communicating details of Xetá culture, which, without his help, will be lost to posterity. However, the atmosphere

is relaxed, and Loureiro appears much more in control of the situation than
in the forest, and there seems to be a positive relationship between him and
Tuca.

3. Contact with the Xavante

Photograph 3: A Tense Handshake (1949)

Attempts to make contact with the Xavante people were marked by hostility
and violence. Two Salesian Catholic priests attempting to make contact with
the Xavante had been killed in 1932, and nearly all members of the first
pacification expedition, led by SPI inspector Genésio Pimentel Barbosa, had
been murdered by the Xavante in 1941, with only a few workers and
interpreters (of the Xerente group) managing to escape as they were far from
the main group of the expedition when the encounter took place (Freire &
Guran 2010:68). The Brazilian media stressed the bravery of the Xavante
and their fierce resistance to outsiders. Five years later, a second team was
organized, now led by sertanista Francisco Meirelles, and when contact was
finally made, and goods were exchanged with the Xavante, led by Apoena,
the media and the state commemorated. The image here is from a later 1949
expedition.

In the photo we see the great tension of the situation. On the SPI side, this anxiety was certainly caused by past violent events and the reputation of the Xavante as hostile and cruel; on the Xavante side, because of its long history of conflicts with the government and deaths from disease and war. Distance is maintained between all the people involved, showing that trust does not yet exist. It should be noted that the interpreter Euvaldo Gomes was not a Xavante; in fact, he is not even indigenous, which may have produced even more distrust and fear among the Xavante.

In western culture the handshake is a symbol of showing long-standing friendship. By holding out empty right hands, strangers could show that they were not holding weapons and were not showing bad will. The up and down motion would also dislodge any knives or daggers that were hidden in the sleeve. Another explanation is that the handshake was a symbol of good faith when taking an oath or promise. When they held hands, people showed that their word was a sacred bond (Andrews 2020:Internet).

As already seen, in Western culture we naturally look at visual material from left to right (Kress and van Leeuwen 2006:179-180), and on the left we usually have the old, already known information, and we look to the right to see the new information. Thus, in most paintings of the Virgin Mary and Baby Jesus, as we shall see in the example in the next chapter, we find the Virgin Mary on the left; she is the living mother, and the newborn Jesus, the new element, is found more to the right; and when there is a saint adoring the baby Jesus, they are normally placed on the right.

In *Art and Visual Perception* Rudolf Arnheim describes the work of several authors who describe the difference between left and right. Heinrich Wölfflin (1941:82-96) points out that the diagonal running from the bottom left to the top right is seen as ascending while the other diagonal running from the top left to the bottom right is seen as descending, so both diagonals start from the left side. And an object on the right side is always heavier. Furthermore, experiments show that when two similar objects appear on each side, the one on the right side always appears larger (in Arnheim 1974:34).

In Chapter 1 we saw Mercedes Gaffron's idea that the observer always identifies with the left side, and what appears there is of greater importance. In a mirror image of a photo, an object in the foreground in an asymmetrical scene appears closer when it is on the left side. In the theatre the audience looks first at those who enter from the left and looks more at the actor who is on the left of a group on stage. Gaffron relates this to dominance of the left side of the cerebral cortex, which contains the higher brain centres for speaking, writing, and reading. So as we are more familiar with what we visualize on the left, we notice more what is on the right, which explains why the objects that appear there are more conspicuous and different. Intensified attention to what is happening on the left side would compensate for this asymmetry, and the eye would spontaneously move from the site of initial attention (left) to the area of more articulated vision (right). The right side is therefore more visible, and the object on that side is heavier; and the left side is more central and the side with which the spectator identifies. Thus, in the theatre the more pleasant characters appear on the left, and the bad characters on the right. It seems easier for a gentleman on the screen to go from left to right than vice versa, and the psychologist H. C. van der Meer observes that we move our heads more quickly from left to right (Arnheim 1974:34-35).

Using these concepts, we can examine *Photograph 3*. We see the Xavante boy on the left, and Euvaldo Gomes, who organizes the action in the photograph, is in the middle and looking tense. And we can also read the photograph as a kind of tribute paid to the boy, Christ being adored by the wise men, the saints, and the shepherds, the interpreter Euvaldo Gomes, representative of the white Brazilian State, showing respect for the Xavante indigenous people, represented by the boy.

Euvaldo Gomes, aware of the problems that occurred in the previous contact with the Xavante, extends his right hand towards the boy, and this gesture could hardly be recognized as a threat. He can't be hiding any gun or knife. The hands meet, and so do the hands of the figure in the background, occupying a central but hidden position in the photo, an Indian, probably a friend or a brother of the boy in the foreground. It looks as if he's turned his head, and we can't see anything of his face, but his right hand enters the four-handed grip. A scene of friendship, perhaps a sign that the Xavante are already entering the western world.

As in *Photo 6* of Chapter 1, *Vissière Translates Alfred Leblanc*, with smiles on all sides, in an atmosphere of success, harmony, and satisfaction, we expect a kind of positive resolution of the photo on the right, perhaps a group of Indians smiling, applauding this new contact, and the friendship found between the whites and the Indians. Indeed, if we cover the figure on the right, we receive a very different impression of the photo. We remember with Arnheim that the right side is the side that has more visual weight in the photo, and new elements here have a greater weight than if they appeared on the left or in the centre. Thus, on the right side the other older indigenous man appears, with an oblique gaze, towards a vanishing point to the left of the camera but with the camera within range of his vision. He is suspicious of the camera and/or something/someone else, challenging it, and/or protecting the boy, and he shocks and baffles us. Is he the father, uncle, grandfather, or guardian of the boy in the foreground? Does he suspect that the camera might be some kind of weapon?

Kress and van Leeuwen emphasize the importance of the angle of the shot. A frontal angle, with the camera looking directly into the photographed person's face, signifies a certain involvement with the subject: "What you see here is part of our world, something with which we are involved"; and the oblique angle says the opposite: the camera, the photographer, the spectator is not part of this world, has no involvement with it (Kress and van Leeuwen 2006:136). Thus the photographer will choose their position: frontal, horizontal, or oblique, and the angle of the oblique position can vary. In an oblique photo everything will appear at an angle, but when the photographer takes a horizontal position, a side view, it is very possible that some element in the photo is facing the camera. Here the photographer was careful not to point the camera directly at the boy's face and chose a horizontal side view of 90 degrees, but it seems that the man on the right side, worried, entered uninvited, thus changing the initial composition of the photo.

Thus *A Tense Handshake* is a highly disturbing photograph. We expect resolution on the right side of the photo, the most conspicuous point, but we only find a problem, a threat. The attempt to open friendly contact with the Xavante is already encountering problems, and we fear for our interpreter, Euvaldo Gomes. Will this expedition be more successful than the last one? What will result from the mistrust of the figure on the right? With hindsight, we know that this expedition was the beginning of the pacification of the Xavante, but the path seems full of risks and dangers.

There is yet another disorienting element: we see that the three male figures all have something white tied around their penises. The boy, thin but healthy, about eleven years old, has it, like the other one who is semi-hidden, and the disturbing figure on the right. We are before the unknown. What is it? It is Barthes' *punctum*. Three small, white triangles stand out from the photograph. We ask: why does this society put these pieces of cloth (would it be cloth or animal skin?) on men's penises? What are they for? To hold the penis? Do they hurt? Are they ornamentation? Do they make erection difficult? Urination? Are they taken off to have sex? To sleep? A little investigation solves the problem, which we found in Maria Cristina Resende dos Campos' MA dissertation, "O Corpo Emana: Elementos da Plástica Corporal Xavante" ["The Body Emanates: Elements of the Xavante Bodily Plastic"] (2007):

Figures 1 and 1a: The Penis Sheath

It is a protector of the male sexual organ, made from palm bud leaflet and consisting of a funnelled fold, with a hole at the tip, to press the foreskin and hide the glans. The protector therefore only covers the end of the sexual organ. Apart from this, they wore no other type of clothing and walked around completely naked. According to Maybury-Lewis (1984), for the rest of their lives they will never be seen without the case, even in the bath. The Xavante considered it extremely disrespectful for a man to be seen without this protection. The only times they took the case off were determined by physiological needs such as urinating and copulating[14] (1984:84-85).

And quoting Maybury-Lewis:

> The use of the penile sheath corresponds to a symbolic affirmation of physiological maturity; the sheath "covers" the penis in an erection position and thus indicates sexual potency while affirming the sexual power to which intrinsically dangerous sexual powers are subjected (1984:156).

4. The Interpreter Vegnon and the Kaingang Pacification

This photo is undated, and the only information provided together with the photo is about Vegnon's background and that he was an interpreter. Chief Kaingang Vegnon was one of the interpreters of the attraction team assembled to work on the pacification of the Kaingang in the west of the state of São Paulo, which took place from 1910 to 1912. The main interpreter of the expedition was Vanuíre, a Kaingang woman known as "the Pacifier" because of her qualities as a mediator and the role she played during the process of contact with the Kaingang. However, despite Vegnon not receiving as much attention as Vanuíre in official reports, he, along with the interpreter Futoio, was responsible for establishing the first successful communication with Chief Vauhin and other hostile Kaingang natives, which resulted in the clarification of very important information for the peace process. Shortly thereafter, on 19 March, 1912 ten Kaingang warriors, including Chief Vauhin, completely unarmed and naked, approached the camp. The visits then increased in frequency, and the SPI began to establish peaceful contacts with the Kaingang in the west of the state of São Paulo (Barbosa 1947; Barbosa and Milton 2021).

We have no information about where the photograph was taken, and in the book *Memória do SPI*, the same photo describes him as a member of the

---

[14] "Trata-se de um protetor do órgão sexual masculino, feito de folíolo de broto de palmeira e que consiste em uma dobradura afunilada, com orifício na ponta, para imprensar o prepúcio e ocultar a glande. O estojo s. cobre, portanto, a extremidade de seu órgão sexual. Além do estojo, eles não usavam nenhum outro tipo de vestimenta e andavam completamente nus. De acordo com Maybury-Lewis (1984), pelo resto da vida, jamais ser.o vistos sem o estojo, mesmo no banho. Os Xavante consideravam extrema falta de decoro um homem ser visto sem essa proteção. As únicas ocasiões em que tiravam o estojo eram determinadas por necessidades fisiológicas como urinar e copular".

Photograph 4: The Protector of the Indians (1910 - 1912)

Terena group, not a Kaingang, wearing a military uniform from the Paraguayan war (Freire 2011:168). However, there is agreement as to the fact that he, Vanuíre, and Futoio were brought from São Paulo, specifically from Aníbal Sodré's farm, where they worked as slaves (Pinheiro 1999:161).

What kind of message does the photo give us? The primitive wattle and daub hut hardly seems to offer protection to the indigenous population. It seems like a rather timid effort on the part of the Brazilian government to offer this support. Vegnon's isolation draws the viewer's attention. Will only Vegnon protect the Kaingang? Indeed, why is he alone? Can we see him as a traitor to his own Kaingang society like the several lone interpreters described in *Framing the Interpreter*? Why is he wearing the uniform? He seems to be proud of wearing it, and, according to Luciana Alves Barbio: "The military uniform was an element of extreme distinction among the indigenous people and was, along with the work at the telegraph posts, an important symbol of equivalence between indigenous people and the military" [15] (Barbio 2005:96, in Melo Santos:Internet). In his biography of Marechal Rondon, Larry Rohter writes that it was an enormous honour for a Bororo native to receive a uniform from Rondon: "His oldest friend in the tribe adopted the name "Cadete" ["Cadet"] in Portuguese, chosen after Rondon gave him a military uniform as a token of appreciation for his work on the Pantanal telegraph line (1900-3)[16] (Rohter 2019: 447). And "Uazá was also given Rondon's own uniform of Major Engineer; and, Tohoerá, his son, the uniform of 2nd Lieutenant of Artillery of Lieutenant Lyra" [17] (Melo Santos:Internet).

However, when we discover that the uniform is that of the Brazilian army in the Paraguayan War (1864-1870), his patriotic pride seems absurd, and we realize that he has received some old leftovers, making him look more like a scarecrow or as if he were going to a costume ball than a real soldier.

---

[15] "A farda militar era um elemento de extrema distinção entre os indígenas e configurava, junto com o trabalho nos postos telegráficos, um importante símbolo da equivalência entre indígenas e militares".

[16] "Seu amigo mais antigo na tribo adotara o nome de Cadete em português, escolhido depois de que Rondon lhe deu um uniforme militar como sinal de agradecimento por seu trabalho na linha telegráfica do Pantanal (1900-3)".

[17] "Ao Uazá foi dado ainda o próprio uniforme de Rondon de Major Engenheiro; e, a Tohoerá, seu filho, o uniforme de 2o Tenente de Artilharia do Tenente Lyra".

Rancho de caçadas encontrado pela comissão no dia 28/6/1911
Êste rancho foi encontrado na data acima, em plena floresta depois de uma
marcha de oito horas, partindo de "Heitor Legrú". No rancho havia sinal de

Photograph 5: Discovery of the Settlement (1911)

This photo gives a very different idea of Vegnon. He is the interpreter guide, or *língua* [tongue], the traditional name he was given in Portuguese, who probably led the SPI group to the discovered ranch. After a long hike he looks tired but has a more confident appearance than in the previous photo. He is no longer wearing his military uniform and appears to be integrated into the group, with a look of tranquility, confidence, and satisfaction at having successfully completed the task in which he was a major participant. He also looks much healthier than in the previous photo, where he seemed quite shrunken in his uniform, and must have been eating much healthier food.

5. Darcy Ribeiro and the Ka'apor

Photograph 6: A Good Day at the Office (1951)

In 1947 the anthropologist Darcy Ribeiro was hired as a "sociological anthropologist" by the SPI, which had created a Studies Section (SE) in 1942, after the dictatorship of Getúlio Vargas decided to invest more heavily and restructure the SPI (Freire 2011:213-214). He wrote *Diários Índios* on two expeditions to the Ka'apor villages located between the Gurupi and Pindaré rivers, on the border of the states of Pará and Maranhão in 1949 and 1951, although they were published only in 1996 and contain revealing comments on SPI interpreters.

Darcy had major problems on the first expedition, from 20 November 1949 to 11 April 1950. Although he had some knowledge of the Ka'apor language (which is a member of the Tupi-Guarani language family, whose varieties were and still are spoken in many areas of Brazil, especially in the south and southeast), he needed an interpreter to help him with details in several areas: genealogy, anthropophagy, censuses, rituals, work routines, myths and traditional stories, crafts, all described in the diaries. Initially, Darcy apparently had no choice and was allocated an SPI employee who spoke the Tembé language (also from the Tupi-Guarani family), who was given the chance to earn extra money for his services. Ideally, the interpreter would have intimate contact with the Ka'apor, be able to get along with them as a sort of liaison interpreter, and extract from them as many stories and legends as possible, together with information about kinship and daily habits, an integral part of the anthropologist's research. He would have know the area well and at times be prepared to work as an interpreter-guide. The work was often quite complex, requiring skills very different from those needed by the interpreter-guide who might come into contact with an isolated group. As the language had no written form, all interpretation would be oral. Darcy's interpreter, Emiliano, a Tembé Indian, who could communicate with the Ka'apor, their language being from the same linguistic family, was a complete failure, restricting the amount of information Darcy could obtain.

In fact, interpreters' work in general was considered of little value. In the 1986 FUNAI job and salary plan, interpreting is at the lowest level:

1) Occupational group I, formed by semi-skilled labour. This category included the functions of field service assistant (four levels), involved with the work of the *sertanista*[18]: indigenous interpreter, river pilot, assistant to the *sertanista*, nursing attendant;
2) Occupational group II, with field administrators and indigenous field administrators (I and II) and the *sertanista*. Their basic basic activities were to be responsible for decisions, participate in the formation of expeditions, "participate in the plans of contacts and entering into indigenous communities" and carry out "attraction, pacification, and settlement services for indigenous groups" (PCS, 1986);

---

[18] The original *sertanistas* in the 17th century in the hinterland of Brazil captured and enslaved Indians in their search for precious metals. Nowadays the *sertanista* is seen more as a guardian of the forest and a link with indigenous groups.

3) This occupational group refers to higher-level positions. It is important to observe that the salary of *sertanista* corresponds to the initial salary range of a higher level[19] (Freire 2005:40).

Later on in the first expedition, Darcy meets João Carvalho, a *sertanista*, whose interpreting skills Darcy admires, and with João Carvalho as a regular interpreter, the second expedition, from 2 August to 9 November 1951, was much more successful. João Carvalho successfully communicated information on many topics to Darcy: he provided details of the geography of the Ka'apor territory (Ribeiro 1996:338, 11 August 1951); he knew indigenous slang (Ribeiro 1996:91, 8 February 1950); he took responsibility for organizing the groups and individuals he would meet and interview (Ribeiro 1996:331, 8 August 1951); and as a result of his skills, the number of myths that Darcy manages to collect and accurately transcribe on this expedition is much higher. The interpreter's duties were not exclusively oral, and João Carvalho protected him from mad dogs chasing a peccary [wild pig] (Ribeiro 1996:553, 20 October 1952).

Darcy emphasizes the important role of his interpreter: "João is our interpreter and in these hours of social receptions his role is as important as

---

[19] 1) Grupo ocupacional I, formado por mão-de-obra semiqualificada. Nesta categoria se enquadravam as funções de auxiliar de serviços de campo (quatro níveis), envolvidos com o trabalho do sertanista: Intérprete indígena, Motorista fluvial, Auxiliar de Sertanista, Atendente de Enfermagem, Piloto de lancha;
2) Grupo ocupacional II, tendo como assistentes técnicos de campo, o técnico de indigenismo (I e II) e o Sertanista. Esses técnicos tinham como atividades básicas responsabilidade de decisão, participar da formação de expedições, "participar nos planos de contatos e penetração nas comunidades indígenas" e executar "serviços de atração, pacificação e aldeamento dos grupos indígenas" (PCS, 1986);
3) Esse grupo ocupacional se refere aos cargos de nível superior. É importante observar que o salário de sertanista corresponde à faixa salarial inicial de nível superior. 1) Grupo ocupacional I, formado por mão-de-obra semiqualificada. Nesta categoria se enquadravam as funções de auxiliar de serviços de campo (quatro níveis), envolvidos com o trabalho do sertanista: Intérprete indígena, Motorista fluvial, Auxiliar de Sertanista, Atendente de Enfermagem, Piloto de lancha;
2) Grupo ocupacional II, tendo como assistentes técnicos de campo, o técnico de indigenismo (I e II) e o Sertanista. Esses técnicos tinham como atividades básicas responsabilidade de decisão, participar da formação de expedições, "participar nos planos de contatos e penetração nas comunidades indígenas" e executar "serviços de atração, pacificação e aldeamento dos grupos indígenas" (PCS, 1986);
3) Esse grupo ocupacional se refere aos cargos de nível superior. É importante observar que o salário de sertanista corresponde à faixa salarial inicial de nível superior.

in those of work, so he has to say a lot to compensate for my silence"[20] (Ribeiro 1996:366, 22 August 1951). He admits his frustration at having to use an interpreter, but thanks João:

> How difficult it is to speak through someone else's mouth, it is only worse to hear with other people's ears. I feel such anguish at listening to the Indian speak at length and only get a laconic translation. But I must not complain, João is the best interpreter I have had, and for him my questions and demands, which he cannot always understand, are much more distressing than his worst interpretations are for me[21].

In the first expedition he mentions the measles outbreaks, which caused the deaths of many indigenous peoples, and the bad conditions he was forced to endure. In Piahú, he is lying in the hammock and writes:

> There's no table around here, nor anything that makes it possible to write. The best thing is to keep swinging in someone else's hammock. Mine is at the captain's house, among all the ill patients. This is, without a doubt, the place where I've felt the smell of humanity the most. What animal stinks more than man?[22] (Ribeiro 1996:180, 4 February 1950)

And in the village of Ianawakú, on the first expedition: "The mosquitoes also torment me, it's torture, my whole body is already scarred. Even bumps in my groin, which after worrying a lot I found to be the result of so many insect bites"[23] (Ribeiro 1996:126, 7 January 1950).

However, on the second expedition, he visits healthier settlements, such as Tapuro Ambir Hechuan, which is free of measles and other diseases, and

---

[20] "João é nosso intérprete e nessas horas de recepções sociais seu papel é tão importante quanto nas de trabalho, de modo que tem de falar muito para compensar meu silêncio".

[21] Como é difícil falar pela boca alheia, só é pior ouvir com ouvidos alheios. Imensa é a angústia de escutar o índio falar longamente e só obter uma tradução lacônica. Mas não devo queixar-me, João é o melhor intérprete que tive, e, para ele minhas perguntas e exigências, que nem sempre pode compreender, são muito mais angustiosas que suas piores interpretações para mim] (Ribeiro 1996:367, 25 August 1951.

[22] "Por aqui não há mesa, nem qualquer coisa que se preste para escrever. O melhor é mesmo ficar embalando em rede alheia. A minha está na casa do capitão, no meio de todos doentes. Este é, sem dúvida, o lugar onde mais senti o cheiro de humanidade. Que bicho federá mais que o homem?"

[23] "Os mosquitos também me atormentam, é um suplício, já tenho todo o corpo marcado. Até ínguas nas virilhas, que depois de muito me preocupar verifiquei serem consequência de tanta picada de insetos".

where conditions seem even idyllic, with a "happy rhythm" ["ritmo feliz"] (Ribeiro 1996:367, 25 August 1951). Darcy and his team were there from 22 to 27 August 1951, and *Photograph 6* is taken during this period by official SPI photographer Heinz Förthmann.

As we have already seen, Darcy often complains about dirt and lack of hygiene, but here he appears clean, shaved, combed, perfumed, and wearing freshly laundered clothes. It also seems that João Carvalho has just shaved. The smiles, particularly Darcy's Cheshire Cat smile, may even look a little fake. Is it the forced smile of someone having an official photo taken? Is it part of Darcy's questioning technique? Or does it come from his contentment at successfully completing his job with the help of João Carvalho?

Darcy stands out in his white linen clothes, which makes him an outsider, probably a way of preserving some scholarly distance from his subjects in the odd setting. He writes in the *Diário* that he does not try to dress like a native although he admires the nakedness of the Indians.

The photo was taken at an oblique angle by a professional photographer, but it looks deliberately unposed and even amateurish, with João Carvalho being partially cropped.

We see a formal interview although we see nothing of the interviewee's face, in a formal roleplaying situation, with tables and benches inside a communal tent. Furnishings were not part of native tents, and it appears that the benches and table were quickly improvised to help Darcy conduct his study and official research and note down details. But what's missing are Darcy's tools of the trade, his notebook, pen or pencil, and his sonographer to record the voices. As seen above, Darcy often complains that he has no place to write. Now he has a table, but he's not writing. So maybe this is a staged photo, just before the actual interview, and Darcy is trying to put the interviewee at ease.

What are other people doing there? Why is the other woman's back turned? Is she next in line to be interviewed? What is she looking at? Why was she included in the photo? Who is the woman behind Darcy's left shoulder? Darcy observes that the Indians were always curious to know what he was doing and would often interrupt him, but this was more the case with the men than the women.

Perhaps their presence is due to the very cramped premises. All natives living under one roof. Maybe they just had nowhere else to go. Darcy mentions the absolute lack of privacy: everyone knows everyone else's business:

> Each person who leaves tells someone, when passing, where they are going. Thus, everyone knows, at any time, where each one is. This means that, even outside the home, they are not only coexisting, but living their whole lives together, which becomes unbearable for people like us[24] (Ribeiro 1996:256, 5 March 1950).

6. The Transformation of the Indian: from Koluizorocê to Major Libânio

The sequence of photos from "Os Brasis e sua Memórias" ["The Brazils and their Memories"] from the biography of Major Libânio Koluizorocê by Rita de Cássia Melo Santos (Melo Santos:Internet) shows the integration of the Indian into western society. The information that follows is taken from that biography. Koluizorocê was a member of the Ariti who played a central role in the conquest, clearing, and consolidation of the state occupation of the backlands of Mato Grosso undertaken by the Strategic Telegraphic Lines Commission from Mato Grosso to Amazonas (CLTEMTA) headed by Cândido Mariano da Silva Rondon between 1906 and 1915 and, therefore, known as the "Rondon Commission". Libânio was Rondon's main guide in the second expedition in 1908, when he carried out the reconnaissance of the region between the Serra do Norte and the Madeira River. On his return to Rio de Janeiro in 1910, Rondon took Libanio, who also accompanied him on the 1913-1914 Rondon-Roosevelt Scientific Expedition, an expedition to survey the course of the Rio da Dúvida ["River of Doubt"] in the Amazon basin, in which the ex-President of the United States, Theodore Roosevelt, took part. The trip to Rio de Janeiro was repeated several times until his last trip to participate in the commemorative events of the centenary of Independence of Brazil in 1922.

---

[24] "Cada pessoa que se afasta conta a alguém, de passagem, para onde vai. Assim, todos sabem, a qualquer tempo, onde está cada um. Isso significa que, mesmo fora da convivência fora da casa, estão não apenas coexistindo, mas convivendo a vida toda, o que acaba por ser insuportável para gente como nós".

Photographs 7, 8, 9: "The Transformation of the Indian"

The guide-interpreters ensured the first communications and facilitated the identification of settlements and trails established by previous populations, in addition to enlisting manpower for all the work necessary for the expeditions, from loading supplies and cargo to clearing the ground and later installing the telegraph posts.

During his second expedition in 1908 Rondon inaugurated the first station, on the left bank of the Juruena River, with 52 men commanded by 2nd Lieutenant José Joaquim Teixeira da Silveira. He aimed to establish the initial posts and then proceed with the settling of the Paresí Indians around these posts. He would take advantage of their labour to maintain the telegraph line, raise cattle, and cultivate crops. It was also during this expedition that the first signs of the presence of the Nambikwara Indians were found.

It was in this expedition that Rondon first began to cross unknown regions occupied by hostile Indians. Now Libânio became his main guide though fellow members of his Ariti tribe refused to venture into the unknown and hostile territory of the Nambikwara. However, the expedition was carried out successfully, and a third expedition was made in 1909.

During the expedition Libanio acted on two major fronts. The first as a scout, looking for information left by the reconnaissance team. And on a second front, Libânio contributed to the research carried out by the scientists who accompanied the expeditions, and through him Alípio de Miranda, a zoologist at the National Museum, received a specimen of Speothos Venaticus Lund, a type of wild dog; it was also through him that Frederico Carlos Heohne, a botanist at the Museu Paulista, came to know the composition of the poison used by the Nambikwara in their hunts; and in 1912, after subsequent expeditions, Libânio gave Edgar Roquette-Pinto (1884-1954) material for the formation of the ethnographic collections of the National Museum.

Roquette-Pinto, in his passage through Utiariti, mentions that Libanio headed many families, who lived happily there. According to the anthropologist, the manioc and maize fields maintained the entire group formed almost exclusively by Paresí, with the exception of two whites (Roquette-Pinto 1935:294, in Melo Santos:Internet).

The last news we have about Libânio is his participation in the commemorative exhibitions of the centenary of the Independence of Brazil, held in Rio de Janeiro, between September 1922 and March 1923. On that occasion, more than 14 countries were gathered, and it was attended by some three million people. Libânio was assigned with a group of indigenous people to present "Zicunati", an Ariti game that consists of heading a ball. Played between two groups of 8 or 10 or 15 players, each led by a chief. The game was called "headball" by former President Roosevelt. Apparently Libânio died on this trip when passing through the state of São Paulo, possibly poisoned by a rancher to whom he refused to teach this game (Machado 1994, in Santos 2022).

Libânio was never in fact a Major in the Brazilian Army. He was called this as a form of respect, just as his father-in-law had received military uniforms years before.

The photographs show the changes of Libânio from the native to the soldier and his integration into Brazilian civilization, perhaps an example of the integration that Rondon proposed for the Brazilian Indians: the first photo is of the Indian Koluizorocê[25], chief of his tribe, still in the "primitive" state, holding spears and nearly naked. In the second photo, we now see Libânio Koluizorocê dressed in clothes of white society, sitting in a chair, a symbol of the western house, acculturated, with no trace of his Indian life. He wears a linen suit, befitting a chief of the tribe. This acculturated Indian would not belong to the lowest layer of Brazilian society. And in the third photo we see Major Libanio Koluizorocê, Marechal Rondon's trusted man, who has an important role in the project to occupy the backlands of Mato Grosso undertaken by the Strategic Telegraph Lines Commission from Mato Grosso to Amazonas (CLTEMTA), headed by Rondon. The uniform he is wearing may be the one he received from Rondon himself, and, with his little moustache, he even physically resembles Rondon, who had Guaná indigenous blood on his father's side, and Terena and Bororo on his mother's side.

In this chapter we have seen photographs of various interpreters and the very different relationships between interpreters and those interpreted. First we saw the shyness and isolation of Tuca, the young Xetá indigenous boy; then the moment of great tension when Euvaldo Gomes tries to shake hands with the young Xavante; in the two photos of Kaingang there is a great contrast between the stiffness in the first photo and the more relaxed body in the second; we saw a photo with the smiles of Darcy and his interpreter João Carvalho, surely because João's interpretation was successful; and finally we saw three photographs that show the complete transformation of the Indian Koluizorocê into the soldier Major Libanio.

---

[25] Rita de Cássia Melo Santos, in her biography of Major Libânio Koluizorocê in "Os Brasis e sua Memórias", uses this picture. Elsewhere, such as the "Povos do Brasil: Paresí site, the same picture has the caption: "Chefe dos Arití-Uaimaré, Uazácuriri-gaçú, Mato Grosso". This warrior also appears in photos of the Paresí we shall look at in Chapter 5. The Museu do Índio does not register it as a photograph of Major Libânio. However, his features are similar to those of Libânio in the other two photos.

CHAPTER 4

INDIGENOUS INTERPRETERS IN BRAZIL

1.1. The First Contact with Civilization

Photograph 1: The First Contact with Civilization

This photograph is part of a 1996 study by photographer Ricardo Beliel when the first contact was made with the Korubo by the FUNAI [Fundação Nacional dos Povos Indígenas] [National Foundation of Indigenous Peoples] Contact Front, led by Sydney Possuelo (1940-), an anthropologist specialized in making contact with isolated groups who was Director of FUNAI between 1991 and 1993.

Beliel describes their approach on 15 October 1996:

> When we entered the village, the Indians who were with us were very nervous because they realized that the Korubo were surrounding us. After

all, we were invading their territory. When Sydney and the other members of the expedition arrived, they encountered the Korubo along the way. There was a moment of apprehension, and then they used the tactic of starting to laugh out loud to break the ice. In a magical and surreal way, the Indians themselves started laughing, approaching and touching us. They macerated leaves with their hands and rubbed them on our bodies and faces. For them we were the ones who were being tamed as they were tired of being hunted in a cowardly way by whites like us[26] (Beliel 2013:Internet).

On the lower left we see Possuelo kneeling, smiling, making contact with the child of a young mother with the characteristic Korubo haircut, and paint splatters on his face and left arm. She is carrying another child, aged two or three, who has paint stains on its face and back; her rounded belly suggests she could be pregnant again. She looks at Possuelo, intent but without any hostility. The fact that Possuelo was allowed to have contact with the women and children showed that the Korubo had a certain trust in these strangers.

On the right there is another young woman, with a hairstyle very similar to the first, also with a young child in her arms and another of three or four years-old on her back. Her hairstyle is similar to the first girl's, and she smiles slightly, but the latter's child, like the child on the back of the first girl, seems somewhat scared.

The interpreter, Binan Tuku Matis, wearing a green T-shirt and blue shorts, talks to two people we can't quite make out, perhaps a man and a boy, without children. The Matis are a group related to the Korubo, with similar languages, and certainly the interpreter would have been able to communicate certain things.

In the background, behind the interpreter, we see an older man, who does not participate in the central scene. Possuelo's left hand appears to be resting on the right leg of a Korubo man whose face we can make out behind the

---

[26] "Quando entramos na aldeia, os índios que estavam conosco ficaram muito nervosos, porque perceberam que os Korubo estavam nos cercando. Afinal de contas nós é que estávamos invadindo o território deles. Quando Sydney e os outros expedicionários chegaram, eles encontraram os Korubo pelo caminho. Houve um momento de muita apreensão, e então eles usaram a tática de começar a rir bem alto para quebrar o gelo. De uma forma mágica e surreal, os próprios índios começaram a dar gargalhadas, se aproximar e a nos tocar. Maceravam folhas com as mãos e esfregavam em nossos corpos e rostos. Para eles, nós é que estávamos sendo amansados, pois estavam cansados de serem caçados covardemente por brancos como nós".

woman in the centre of the photo. The impression we have is that Possuelo (and Beliel, the photographer) do not want this man to appear in the photo, as he could spoil a portrait that is essentially a tribute to women and children.

Formally, this is a picture that closely resembles many traditional religious paintings, the adoration of shepherds or wise men, or saints. *St. Francis Kneeling before the Virgin and Child* by Jacopo Bassano (1600) is one of many examples, showing St. Francis kneeling in respect and humility, demonstrating adoration of the mother of Jesus and baby Jesus. And here, in Beliel's photo, the adoration of the two mothers and their two children!

Painting 1: St. Francis Kneeling before the Virgin and Child

The photo also pays homage to the fertility of women. The four children look healthy, and maybe another one will arrive soon. This sense of fertility is emphasized by the two dominant colours, green and brown, the colours of nature.

We see hands, heads and legs everywhere, those of Possuelo, the interpreter, the mothers and the children, with a total of fourteen hands, eleven heads and fourteen legs. But the central focus is on Possuelo's right hand, reaching out to hold the hand of the baby Korubo, which reminds us of God's hand reaching out to touch Adam's in Michelangelo's *The Creation of Adam* (c.1508-1512).

Painting 2: The Creation of Adam

Dorothea Lange, the famous American photographer of peasants in Oklahoma during the depression of the 1930s, emphasizes the importance of hands in photography:

> And it is in the hands – rather than the head or the eyes – that this mental activity is most vividly apparent. Even when the heart has all but shut down the fingers are still fidgety with life; when passive they have the troubled serenity of dreams which continue unabated and unsatisfied (in Dyer 2007).

The worn fingertips and the deep lines of his palm, showing the painful work of picking cotton and a lifetime of manual labour dominate one of Lange's well-known photos (Lange:Internet)

We can also mention a number of photos of President Lula from his 2006 re-election campaign. In "Lula's propaganda boy says his life has improved in recent years" ["Garoto-símbolo da reeleição de Lula diz que sua vida mudou para melhor nos últimos anos"] (Boteko Vermelho 2010:Internet), Lula is in a humbling position on the left side of the photo, drenched in sweat, his bodyguards behind him but allowing a seven-year-old black boy to climb on the back of a man, probably his father, and, smiling and healthy, reaches out his own hand to caress the beard of Lula, whose hands are being grabbed by the man holding the boy and several other people.

In "Lula carried by the people after speech in São Bernardo" ["Lula é carregado pelo povo após discurso em São Bernardo"] (2018) taken at his farewell in São Bernardo do Campo on 8 April 2018, the day Lula would go to jail in Curitiba to serve a nine-year sentence for corruption, he was surrounded by a sea of hands and arms, forming a kind of whirlpool.(Proner 2018:Internet).

"Lula and the hands that make Brazil" ["Lula e as mãos que fazem o Brasil"] (Carvalho 2019:Internet) is a poem and collection of photos showing multitudes of hands of all ages, all colours, men and women, reaching out to touch, grab, caress, and embrace Lula. Also prominent is Lula's left hand, minus its little finger, lost in an accident when he was a metal worker. The hands speak: some showing a lifetime of manual labour, others less so; one hand holds a cross, another has a bracelet, the arm has a small tattoo, but what they all have in common is the affection and love they have for Lula.

As mentioned in previous chapters, Arnheim (1988), and Kreuss & Van Leuwen emphasize the importance of the left side of a picture, especially in the West, where we read from left to right. For Kreuss & van Leuwen (2006) we Westerners, accustomed to reading in that direction, enter a frame from the left. Thus we entered the picture together with Possuelo and Binan Tuku Matis. Their path into the unknown is also ours. We discover the Korubo with them.

We can also analyze the vectors, the energy forces, present in the photograph, representing this first encounter with the Korubo. The vectors coming from Possuelo and Binan Tuku Matis and the indigenous women are directed towards the figure of the central baby, the equivalent of the Baby Jesus in the manger in the paintings of the Nativity (Kress and van Leeuwen:2006).

Photograph 1a: The First Contact with Civilization with the vectors

Beliel took the photo from a horizontal angle, as if he were witnessing a theatrical scene in which he had no involvement. Obviously, in this first contact with the Korubo he wants to be as discreet as possible. He also mentions that he didn't use a camera with a large lens because it could have been mistaken for a gun; and neither did he want to point the camera directly at the faces of the Korubo (Beliel 2013).

Does this picture idealize the Korubo as the Edenic untouched and unspoilt tribe? Other photographs by Beliel of the Korubo, easily found on the Internet, are less romantic and more documental, but the religious connotations of the photograph chosen for this chapter certainly add to this element. Barbara Arisi warns that the Brazilian government policy of isolation for certain groups such as the Korubo contains certain contradictions in its attempt to preserve such groups untouched by "civilization" (Arisi 2018:Internet). Sebastião Salgado's 2017 photographic essay on the Korubo, 21 years after Beliel's picture was taken, may also help to further this myth (Serva 2017), which will be dispelled by the following section.

1.2. The Korubo

Following the website of Povos Indígenas Brasil (Povos Indígenas Brasil, Korubo:Internet), we can look into the background of the Korubo.

The Korubo, also known as "índios caceteiros" [war club Indians"] because of their clubs, live in the region where the Ituí and Itaquaí rivers meet in the Javari valley. Most of this population of more than 200 people still live in isolation, moving between the Ituí, Coari, and Branco rivers.

The first advances in contact with the Korubo began in August 1996, when the village was located in which Ricardo Beliel's photo essay was taken. The team, headed by the sertanista Sydney Possuelo, towed the Jacurapá boat, which would be used as a checkpoint, to the mouth of the Ituí River. Another boat, the Waiká, was the support point in the incursions above the waterfall, especially to inspect the attraction tent, which was a precarious straw hut, built by the members of the Contact Front, on the edge of the Ituí River, close to the path leading to the village. This is where the gifts used in the attraction were placed.

Sydney Possuelo describes the location of the village and the path used by the contact team:

> We entered the clearing at a distance of 50 to 60 metres from the village, separated by a patch of forest of approximately 30 metres. The Indians responded to our songs, they spoke a lot, but they didn't show themselves. We estimated that most of the indigenous people were hunting or wandering around the region. We stayed in the field for no more than 45 minutes. After leaving some gifts we returned to the boat, opening a path along which we hoped that they would visit us. During the two nights on the hike we were surrounded by Indians. They imitated various animals and beat their clubs on the ground, even throwing sticks at the camp. Without major incidents, we returned. We have now made a physical connection between them and us[27] (Povos Indígenas Brasil, Korubo:Internet).

---

[27] "Entramos na roça à distância de mais ou menos 50 a 60 metros da aldeia, separadas por uma nesga de floresta de aproximadamente 30 metros. Os índios responderam aos nossos cantos, falaram muito, porém não se mostraram. Calculamos que a maior parte dos indígenas estava em caçada ou perambulando pela região. Ficamos na roça não mais que 45 minutos. Após deixarmos alguns presentes, efetuamos o retorno ao barco, abrindo uma picada pela qual esperamos que nos visitassem. Durante as duas noites na caminhada, fomos cercados pelos índios. Eles

On 29 August 1996 members of the Contact Front did not find the gifts they had placed in the hut. In response they left more gifts and a bunch of bananas in place of the flour that had been refused by the Indians.

On 15 October 1996 the team established contact with the group, and it was during this contact that Beliel's photo was taken. At the time the group consisted of eighteen people, four women, six men, six boys and two girls. But the positive image of healthy Indians that we see in the photo belies the reality. In early 1998 a man and two boys died after contracting malaria.

And other problems ensued. Ten months of work after the contact, with more than thirty visits made by the Korubo to the base of the Contact Front, members of the Front were attacked by the Indians, who killed the worker Raimundo Batista Magalhães, known as Sobral, with a club. Among the various interpretations of the episode, the most plausible indicates that the reason for the conflict was a canvas taken by the Indians to build a hut. Sobral went to claim ownership of it, and when he picked it up he ended up destroying the hut, whose roof had been made with the canvas.

The last attack undertaken by the Korubo was in 2001, when three loggers were killed on the Quixito River. The attack took place during the period when the Frente de Proteção Etno-Ambiental [Ethno-Environmental Protection Front] was opening a clearing for the construction of the second Surveillance and Protection Post, very close to the place the Korubo had made their attack.

In recent years, contact with non-Indians has been restricted to meetings with members of the Front, especially those who are part of the team at the post at the confluence of the Itaquaí and Ituí rivers. In addition, on occasions Indians have been taken to towns for health treatment.

Leão Serva, who visited the Javari Valley together with photographer Sebastião Salgado in 2017, emphasizes the need for more efficient FUNAI stations to impede illegal loggers, rubber tappers, and drug traffickers, and mentions the incident with the Matis in 2015, when, in retaliation for the killing of two Matis by the Korubo, thinking they were killing invading whites, the Matis killed nine out of a group of thirty, greatly fragilizing the

---

imitavam vários animais e batiam suas bordunas no chão, chegando a jogar paus no acampamento. Sem maiores incidentes, regressamos. Temos agora concluída uma ligação física entre nós".

group and making them dependent on outside help from FUNAI (Serva 2017).

2. How interesting it is to be an interpreter!

Photograph 5: How interesting it is to be an interpreter!

Photograph 5: How Interesting It Is to be an Interpreter! contrasts in many ways with *Photograph 1: The First Contact with Civilization*. First of all, it's a snapshot, an informal, instantaneous photograph taken by an amateur photographer, probably with a cheap camera or an iphone. Wikipedia gives a good idea of the qualities of the snapshot.

> Snapshots can be technically "imperfect" or amateurish: poorly framed or composed, out of focus, and/or inappropriately lighted by flash. Automated settings in consumer cameras have helped to obtain a technologically balanced quality in snapshots. Use of such settings can reveal the lack of expert choices that would entail more control of the focus point and shallower depth of field to achieve more pleasing images by making the subject stand out against a blurred background (Snapshot (photography):Wikipedia).

While *Photo 1* was taken by an award-winning photojournalist, *Photo 5* is formally fuzzy. Is it a photograph of the interpreters in a booth in the background, or a photograph of the Indians in front of the booth, or both? It is actually a photo of the APIC interpreters, hired by the 2012 Rio+20 Earth Summit. The People's Summit was a parallel event to Rio+20, organized by civil society entities and social movements from a number of countries. The event took place between June 15 and 22 2012 at Aterro do Flamengo in Rio de Janeiro, with the aim of discussing the causes of the socio-environmental crisis, presenting practical solutions, and strengthening social movements inside and outside Brazil.

I discovered the photo on the Professional Association of Congress Interpreters (APIC) website in 2018. Unfortunately, it is no longer there. There was no caption of any kind, neither about the indigenous people attending the lecture nor about the names of the interpreters who, after contacting APIC, I found out to be Heriberto Rayón and Vivian Delgado, with two booths: English and Portuguese, and Spanish and Portuguese.

Many snapshots are taken to mark an event (Zuronskis 2013); snapshots are usually "indexical", i.e. the primary function of the photo is to show that the APIC interpreters were there at the Rio+20 Earth Summit (Zuronskis 2013), and our photo is a good example, showing an interpretation session a little different than to the usual: instead of wearing a suit and tie, the audience has their skin painted!

So my initial analysis contrasted *Photo 1* with *Photo 2*, emphasizing indigenous people as ornamental in the photo, asking the following questions: Why are they there? What kind of participation do they have in the congress? Why are they painted? Why are they carrying sticks? Are they going to talk? To dance? Sing?

The contrast with *Photograph 1* can be seen from the following binary terms (*Photograph 1* always opposing *Photograph 6*):

Table 1: ***Photograph 1: The First Contact with Civilization* vs *Photograph 5: How interesting it is to be an interpreter!***

Photograph 1: The First Contact with Civilization	*Photograph 5: How interesting it is to be an Interpreter!*
Empowerment	Lack of empowerment
Empathy of photographer with Indians	Lack of empathy of the photographer with Indians
Contact	Lack of contact
Integration	Lack of integration
Smiling	Serious
Accompanying interpreter	Booth interpreters
Indigenous interpreter	Non-indigenous interpreters
Portuguese-Metis-Korubo	Spanish-Portuguese-English
Women and children	Adults
Reaching out – holding hand out	Everyone contained
Daily life	Special occasion
Human contact	Exotic image
Indians centre stage	Indians on the margin
Creative and artistic photo	Snapshot

How interesting it is to be an interpreter! is a photograph that shows the variety of the interpreter's life. Among other photographs of APIC interpreters working on farms, in hospitals, at conference tables and with recognized figures such as Nelson Mandela, one photo catches our attention:

in front of your interpreting booth there may be exotic creatures like these Brazilian Indians. How interesting it is to be an interpreter!

If *The First Contact with Civilization* shows great empathy with the Korubo, exalting and celebrating them, here the indigenous people are seen as exotic animals inside a cage, although the actual cage in the photo, the interpreters' booth, is the normal work site of the APIC congress interpreters. The interpreters, although unnamed on the website, are the focus of the photo. The Indians, all of whom are also unnamed on the site, are mere decoration, the exotic fauna outside. They are restrained, serious and silent. The interpreters are probably interpreting into English and Spanish for foreign visitors.

My view of the photo changed in May 2018 when I presented it in a lecture at the Philology Week at the Universidade de São Paulo. One of the members of the audience, Helena Barbosa, pointed out that the Indian with the red bands is Megaron, or Megaron-ti, the Megaron Txukarramãe, former director of the Xingu Park from 1985 to 1989 and translator and interpreter of Raoni, leader of the Kayapó, who is also Megaron's father-in-law. Until that moment I had looked at the photo through the lens and through the APIC website, which had erased the identity of the Indians. Now I had another vision.

3. Raoni and Megaron

Photograph 6: Raoni and Megaron, also from the Rio+20 Earth Summit, is a photo that appears on the *Diálogos do Sul* website, showing the indigenous people in a completely different situation, with Megaron translating Raoni's speech, accompanied by the following text by Eliane Potiguara:

> Megaron, Raoni's nephew, endlessly contemplated the sky and with his eagle eyes penetrated the universe as someone looking for the right and focal point, the definition of an answer to the social, political, ethnic and existential problems of the Indigenous Peoples linked by the line of life and the line of clans of the Xingu people purposely chosen by the Universe to bring about the changes that need to take place on Earth[28] (Diálogos do Sul: Internet).

---

[28] "Megaron, sobrinho de Raoni, contemplava infinitamente o céu e com seus olhos de águia penetrava o universo como quem busca o ponto certo e focal, a definição de uma resposta aos problemas sociais, políticos, étnicos e existenciais dos Povos Indígenas atrelados pela linha da vida e a linha dos clãs do povo xinguano escolhido propositalmente pelo Universo para fazer acontecer as mudanças que precisam acontecer na Terra".

Photograph 6: Raoni and Megaron

They are on stage, confident, obviously at ease speaking in public, demonstrating, unlike the passivity of the Indians in *Photograph 5*, a protagonism and dominance. While *Photograph 5* uses an oblique angle, not facing the Indians, avoiding their eyes, not trying to make any kind of involvement in their world (Kress and van Leeuwen 2006:136), *Photograph 6* looks almost directly at Raoni and Megaron, who show no aversion to the camera. Quite the contrary, they are accustomed to speaking in public, disseminating their ideas, and so the more cameras the better.

Helena Barbosa commented to me by e-mail on Megaron's consecutive translation style (and that of other indigenous interpreters who are also leaders in their communities): the translation is not so "accurate" in terms of the original content as it may include explanations and add elements; this sense, it seems, the interpreter rarely translates "to the letter". Megaron, for example, in addition to being the interpreter of Chief Raoni, is also an important Kayapó leader, and thus his opinion and presence have great value, not only semantically, but also symbolically. Depending on the occasion, he makes personal considerations at the time of interpretation, demonstrating a certain freedom in the transposition of speech. Raoni became a leading defender of the environment and indigenous rights, known around the world after visits to Europe.

Since 1964, when he met King Leopold III of Belgium, who was on an expedition in the protected indigenous reserves in Mato Grosso, Raoni has made many international contacts. In 1978 he was the subject of a documentary entitled *Raoni*, narrated by Marlon Brando (Dutilleux and Saldanha 1978). But it was after meeting the singer and musician Sting in the Xingu Indigenous Park in 1987 that Raoni achieved international fame, participating in projects to demarcate Kayapo territories, opposing the Belo Monte dam project and, on visits to several other countries, especially in Europe, spreading their message.

Raoni almost always appears wearing a headdress with yellow feathers and Kayapo earrings and necklaces. He is instantly recognizable by his traditional *botoque*, which stretches out his lower lip and which he displays with great pride.

We can compare *Photo 5: How interesting it is to be an interpreter!* with *Photo 6: Raoni and Megaron.*

Table 2: *Photograph 5: How interesting it is to be an interpreter!* vs *Photograph 6: Raoni and Megaron.*

Photograph 5: How interesting it is to be an interpreter!	*Photograph 6: Raoni and Megaron*
Lack of empowerment	Indians on centre stage
Lack of empathy of photographer	Respectful photographer
Snapshot	Journalistic photo
Lack of integrations of Indians	Indians integrated in a central position
Booth simultaneous interpreters	Megaron as consecutive interpreter
Non-Indian interpreters	Megaron - indigenous interpreter
Portuguese-English (probably)	Kayapó - Portuguese
All are contained	Strong and angry speech

Exotic image for the site	Report for the online report
Indians out of the action, marginal	Indians centre stage
No label or information on the photo	Lots of information about the photo
Oblique angle	Frontal angle

And to conclude this chapter, a third table can compare *Photographs 1, 5, and 6*.

Photograph 1: The First Contact with Civilization	*Photograph 5: How interesting it is to be an interpreter!*	*Photograph 6: Raoni and Megaron*
Artistic and professional photo	Snapshot	Journalistic photo
Horizontal angle	Oblique angle	Frontal angle
Isolated Indians	Indians as curiosities	Indians dominating
Photo full of emotion and symbols	Photo on work of interpreters	Photo showing protest for rights
Interpreter making contact	Simultaneous interpretation	Consecutive interpretation
Hands making contact	Hands inactive	Hand protesting

CHAPTER 5

CHANG AND THE YOUNG WELL-DRESSED INDIGENOUS INTERPRETER

In the preceding chapters we examined photographs about which we had a certain amount of information: relatively little in the case of Arnold Vissière and Julius Meyer in Chapters 1 and 2, much more about Tuca and Megaron in Chapters 3 and 4. And in the final three chapters we shall read about the contradictory but relatively well-documented accounts of the life of John Brown, and the well-detailed careers of Vernon Walters and Viktor Sukhodrev.

But this chapter is something of a leap into the unknown. We have little information on the two photos to be analyzed: that of Chang, the interpreter of the Scottish photographer, John Thomson, on his journey up the Yangtze in 1870, whom we only know through Thomson's comments; and a very well-dressed young indigenous interpreter, apparently an interpreter of Marechal Rondon during expeditions in Amapá in 1914.

1. *Línguas* and *Jurubaças*

In the article "Bolseiros, Lançados, Línguas, Jurubaças and Other Interpreters of Portuguese in Macau and Africa in the Early Modern Period" (Milton 2024) I describe the characteristics of various types of interpreters in the Portuguese colonies. What most of them have in common is that they were usually both guides and interpreters, often coming from the lower echelons of society and mostly illiterate.

In China interpreters almost always came from the lower classes due to the lack of trust in people who spoke foreign languages: speaking another language was a way of betraying the purity of their nation and dirtying its soul: "It was suspected that the soul of the interpreter had been corrupted"[29] (Couto 2011). Thus interpreters were seen as traitors with corrupted souls, and many had even become Christians (Smith 2005). In Macau, for example, interpreters were often from poor backgrounds, had studied at

---

[29] "Suspeitava-se que a alma do intérprete havia sido corrompida".

Catholic schools, and had managed to climb the social ladder by adopting Portuguese names, wearing European clothes and following Portuguese customs (Paiva 2011).

Interpreters frequently existed on the margins of society. They were often dishonest, illiterate, mercenaries, criminals, bastards who had learned the language since birth or through contact with foreigners. In many cases, in the Portuguese empire, criminals were abandoned on the coast of Africa or Brazil and left to fend for themselves. Those who had the luck and skills to survive might find themselves working in commerce between Portugal and the colonies (Silva-Reis, 2018; Silva-Reis and Milton 2019).

Dejanirah Couto writes:

> Former renegades and captives, natives and converted slaves, Jews and new Christians, adventurers and convicts formed an important contingent of a specific category inside the frontier society of the Portuguese empire: that of the interpreters or *línguas* [tongues] as they became to generally be known in the Portuguese empire[30] (Couto 2011).

Línguas often had links to espionage missions and secret negotiations. In Macau the *língua* was responsible for the census and surveillance of the Chinese population in the city and the importation of rice from Canton. However, one of the most important functions was obtaining information, that is, espionage, which involved hiring a chain of informants in Canton, who were placed in strategic positions (Campos 2008:117).

Pay was low and infrequent, work was sporadic, and therefore many interpreters earned most of their money in a variety of ways including bribes, favours, kickbacks, and adding value. And, when accompanying a delegation, there was always the possibility of doing some business or smuggling, as well as receiving

> small donations according to Oriental custom offered by kings or local potentates they visited. Gaspar Martins, interpreter for the delegation of Fernão Gomes de Lemos to Shah Isma'il, received one hundred and fifty cruzados in this manner, as did the clerk Gil Simões[31] (Castanheda, I/III, chap. XLVII: 845, quoted in Couto, 2003:6).

---

[30] "Ex-renegados e prisioneiros, escravos nativos e convertidos, judeus e novos cristãos, aventureiros e degredados formavam um importante contingente duma categoria específica da sociedade dedicada à expansão da fronteira no império português".

[31] "pequenas doações de acordo com o costume oriental, oferecidas por reis ou potentados locais que eles visitavam. Gaspar Martins, intérprete da delegação de Fernão Gomes de Lemos a Shah Isma'il, recebeu 150 cruzados dessa maneira".

And the work was often dangerous. Interpreters often received punishments. The Portuguese were detested by the Chinese Empire, largely because of frequent attacks by Portuguese pirates. When the expedition of apothecary Tomé Pires (c.1465-c.1540), resident of Malacca, today part of Malaysia, entered China in 1516 with some five *jurubaças*, minor interpreters, the main interpreter, the Arab Hoja Asan, died as result of an illness, and the other four *jurubaças* were beheaded, possibly for carrying false documents, and then their wives were sold into slavery. Tomé Pires never got an audience with the Emperor (Letters from the Captives of Canton, in Paiva, 2008, and Paul Pelliot, "Le Hoja et le Sayyid Husain de l'histoire des Ming", in Loureiro 1992, in Paiva 2008).

The following drawing, "Punishing an Interpreter, from 1801", by George Henry Mason (1770-1851), an officer in the British army in China, is of interest, and the fact that the interpreter was included as the only profession, in addition to that of boatman, to have a special punishment, shows that such punishments were not rare and that the profession was considered suspect.

Figure 1: Punishing an Interpreter
A large piece of bamboo cane is placed behind his knees; this is trampled upon by two men, one standing on each end, and who convey more or less pain, as they approach to, or recede from, his person. A punishment, decreed against interpreters, detected of wilful misinterpretation (Mason 1801).

Photograph 1: Chang, the Interpreter

2. Chang, John Thomson's Interpreter

John Thomson (1837-1921), born in Edinburgh, is well known for his photographic portraits of both the rich and poor of London (John Thomson (Photographer): Internet) and, before achieving this fame, he was the first Western photographer to travel widely through part of the interior of China. He first arrived in Asia in 1862, and, based in Singapore, he explored and photographed sites in Ceylon (now Sri Lanka), South India, Cambodia, Siam (now Thailand), and Indochina (now Vietnam). He took his photographs of China between 1868 and 1872 during a four-year stay in Hong Kong, from where he toured various parts of China, and his journeys took him away from the relative comfort and safety of the coastal treaty ports, which had been mostly set up less than a decade earlier upon the conclusion of the Second Opium War. On returning to the UK, he became one of the pioneers of photojournalism, photographing the lives of the poor on London's streets in the 1870s. In the earlier 1900s he became a society photographer and photographed London's rich and famous, even gaining a Royal Warrant in 1881 (MONOVISIONS 2016).

On his expeditions into remote areas of China, carrying his heavy and bulky wooden camera, many large, fragile glass plates, and potentially explosive chemicals, he relied on guides, interpreters, and local labour. These circumstances gave him opportunities for careful study of the habits, character, and social interactions of the native peoples. The expedition of Thomson to the Yangtze River, undertaken in conjunction with "two American gentlemen", was equally well equipped. On this expedition, beginning on 20 January 1870, he hired a translator, Chang and two boat crews to take him along the river. The photograph, *The Interpreter, Chang, from Hankow to Wu-Shan Gorge on the Upper Yangtsze, China*, comes from *Illustrations of China and Its People: A series of two hundred photographs with letterpress descriptive of the places and people represented*, in four volumes, with photos by Thomson, and the narrative of the journalist Adolphe Smith, published in 1873-1874, now available on the Internet.

In the photo of Chang, we see his self-confidence, mastery of the situation, perhaps even his trickery. Roland Barthes writes about the "air" of a photo: "that exorbitant thing which induces from body to soul – animula, little individual soul, good in one person, bad in another. [...] is a kind of intractable supplement of identity [...] the air expresses the subject [...] Perhaps the air is ultimately something moral, mysteriously contributing to the face the reflection of a life value?" (Barthes 1981:109-110) And he gives the example of Richard Avedon's photo of Philip Randolph (1976), an American trade unionist, in which Barthes reads his "goodness".

> Thus the air is the luminous shadow which accompanies the body; and if the photograph fails to show this air, then the body moves without a shadow, and once this shadow is severed, as in the myth of the Woman without a Shadow, there remains no more than a sterile body. It is by this tenuous umbilical cord that the photographer gives life; if he cannot, either by lack of talent or bad luck, supply the transparent soul its bright shadow, the subject dies forever (Barthes 1981:110).

A photo without this "air" is a mere effigy, as could have happened with all the photos that were taken of himself. There is also the risk of never having the "air", the "true image" of the person we love. This risk can be "lacerating" (Barthes 1981:110).

And it is the "look" that manages to give this "air", it is the perception, the thought "seems held back by something interior"; it is "an action of thought without thought, an aim without a target (Barthes 1981:111), and this is what gives the "air". Without looking we simply have attention, it is a *noesis*, the capture of the object by consciousness, but without the *noeme*, the essence. And Barthes gives the example of a photo by André Kertész (1928) of a poor boy carrying a newborn puppy, looking at the camera with sad, jealous, and fearful eyes, a look that Barthes considers "pitiful, lacerating tenderness". For Barthes "he *retains* within himself his love and his fear; that is the Look" (Barthes 1981:112).

And it is this quality that seems to be the "fate" of photography, which makes Barthes believe that he has discovered "the total truth of photography", bringing together the act of capturing the past (*"that-has-been"*) and the truth ("there-she-is!"), bringing a series of emotions such as love, compassion, grief, enthusiasm, desire, even approaching madness, in Barthes' terms, joining "la verité folie" (Barthes 1981:113).

Chang, the Interpreter is a beautiful theatrical picture. We enter the photo from the left (Arnheim 1988:47). It is as if the boatman on the left side is introducing the great star Chang, together with his co-star, who are emerging from the backstage behind. This stage has a double frame: the door frame of the boat cabin behind and on the right the mast; higher up, the beam that forms part of the boat's structure. We arrive at the hub, the centre of the photograph, dominated by Chang, who seems like a figure of authority. In the Western tradition, the central position "has been used to give perceivable expression to the divine or some other exalted power" (Arnheim 1988:109). Chang has an air of disdain and superiority, a command of the situation like someone accustomed to giving orders to others with a certain arrogance. His right hand touches his waist nonchalantly, and his left hand holds a loop of Buddhist prayer beads, a *japamala*, giving him a certain spirituality. He dominates the centre of the photo, looking at a point in the distance, the vanishing point (Arnheim 1974:286-287), whither the eyes of the other boatmen also point. The boat's ropes, two circular coils and two more irregular skeins, help to frame Chang and make the photo a kind of *studium* of the boatman/interpreter – or *lingua* – competent, convinced, pretentious, a little roguish. Did John Thomson manage to capture this "air", "interiority", the *"noema"*, the "essence" of Chang?

Indeed, the photograph is meticulously framed. Elizabeth Edwards stresses the importance of the frame in photographs:

Frame is central to the idea of performativity or theatricality. It has become, of course, one of the central tropes of post-modern thinking, pointing to the multiple layering of constructedness of texts. [...] concern with frame and its metaphorical density has always been active within the discourse of photography, for it is intrinsic to the medium. The frame was one of the five characteristics of the photograph defined by John Szarkowski in his famous catalogue for *The Photographer's Eye* at the Museum of Modern Art in New York: "The central act of photography is the act of choosing and eliminating, it forces a concentration on the picture's edge, the line that separates 'in' from 'out' and on the shapes created by it" (Szarkowski 1966:9) (Edwards 2001:18-19).

The stern of the boat is our stage. The boatman broad on the port quarter presents the star of the show, who is demonstrating his bravado and learnedness as he holds his *japamala* and is supported by his younger sidekick. The ropes, the tools of his trade, frame and highlight Chang's glow and radiance.

3. Questionnaire on Chang

I wanted to see if my impressions of Chang were similar to those of other people. I therefore designed a questionnaire that was sent in Portuguese in May 2020 by e-mail to postgraduate students at Faculty of Philosophy, Letters and Human Sciences of the Universidade de São Paulo (USP), many taking the "Translation and Adaptation" course in the first semestre of 2020, and others from courses taught in previous years. Twenty responses were received. The research did not intend to be of any kind of scientific nature, merely trying to obtain the respondents' impressions of this photo and the photo of the young indigenous interpreter we shall shortly see. The respondents were not asked to read about photography of theory of photography.

1. Where do you think the photograph was taken?

2. When do you think the photo was taken?

3. Who did the interpreter work for?

4. Does the interpreter give the impression of being a good professional? Why?

5. What might the interpreter's responsibilities have been?

6. Describe the interpreter in your own words.

7. Who are the other people?

8. What might they be they looking at?

9. What other elements of the photograph are of interest?[32]

3.1. Discussion of the Questionnaire Results

The replies to the first question mostly recognized that the location was in Asia, more precisely China, and Hong Kong was also one answer. It was also suggested that interpreters would be employed by the English. Two answers suggested it was Mongolia, perhaps Chang's features suggesting he was Mongolian. One response saying that the mentioned the First Opium War (1839-1842), at the very beginning of the invention of photography. Two responses suggested they were from Central America, and one from North Africa or an Arab country.

In Question 2, the majority, thirteen of the twenty answers, said that the photo was from the 19th century or the beginning of the 20th century. One said it was from the 1920s, one the 1960s, and one as late as the 1980s.

In Question 3 two people responded that the interpreters worked for a trader and two others that the interpreters that they worked with fishing and security. Others thought that the interpreters were military personnel. Other responses suggested that they worked for an engineering company, for the photographer himself (correct!), and one maintained the idea of the Hong Kong connection, and that they worked for the British government, and also that they worked for a king or emperor. Interestingly, some respondents were unable to recognize who the interpreter was. One person assumed that the interpreter was the figure on the left, who was not Oriental and worked for Orientals. This idea was shared by another respondent suggesting that the contractor was Chang.

---

[32] "[1. Onde você acha que a fotografia foi tirada? / 2. Quando você acha que a foto foi tirada? 3. Para quem trabalhava o intérprete? / 4. O intérprete dá a impressão de ser um bom profissional? Por quê? / 5. Quais podiam ter sido as responsabilidades do intérprete? / 6. Descreva o intérprete com suas palavras. / 7. Quem são as outras pessoas? / 8. O que ser. que eles estão olhando? / 9. Quais outros elementos da fotografia são de interesse?

Once again, in Question 4 some respondents failed to realize exactly who the interpreter was. Perhaps accustomed to seeing the interpreter in a secondary position, in the background or at the side of a photo, they thought the interpreter was the figure on the left, on the edge of the boat.

Most had a positive impression of Chang. One response emphasizing the courage of the interpreter and another acknowledged that the interpreter was of a higher class than the others. One respondent thought it might be a staged photo. However, a number mentioned that they could not recognize good or bad qualities just by looking at the photograph.

The answers to Question 5 depended on the segment in which they placed the boatmen. They were seen as some kind of police patrol or workers for the British. It was also mentioned that the interpreter's task was to establish contact between the military and the subjugated or that he could be the guide of an expedition.

There were also responses to the effect that they were traders doing the work of translating dialogues between traders and helping to negotiate commercial transactions or appease possible conflicts of interest. Their work might be diplomatic, or communication. One person correctly suggested that the interpreter had been hired to facilitate the photographer's travel.

The function of the interpreter was also seen as very diverse. In addition to translating, he carried out work such as loading, transport and repairs and acted as a guide when traveling through unknown territories, using his local knowledge.

Question 6 elicited both negative impressions such as tiredness and fear and also more positive impressions such as self-confidence, fearlessness, and adaptability.

3.2. Conclusions about the questionnaire

Initially the biggest surprise was the fact that several of the respondents thought the boatman on the left side of the photo was the interpreter and, because of his moustache, thought he was not Chinese. To me it was obvious that Chang was the figure in the foreground but, as has been said, the boatman is occupying a lateral, not central, position normally recognized as being appropriate for the performer.

Other results were more predictable. There is a physical similarity between Chang and the boy coming out of the door, so the suggestion of kinship was predictable. The quality of the photograph is very good; So it was no real surprise that the photo was thought to be from the 20th century. The prominent position of the ropes and gear on the right gives the impression that they were in a fishing boat. Some shared my impression of Chang's haughtiness and self-confidence. But not a single respondent shared my enthusiasm for the beauty of the photo and the way it managed to convey certain elements of Chang's character, or at least my image of Chang.

4. More on Chang

My idealization of Chang was abruptly brought down to earth when reading Thomson's account of the voyage after seeing the photograph. For Thomson Chang was of questionable value. As an interpreter he was of little help, as no one understood his dialect, and Thomson found that his Hanian men, who spoke the Kwang-tung dialect and Malay fluently, were of much more use. But Chang had influence with the Chinese boatmen and authorities with whom they had to deal although this came at a certain cost to the stock of wine and cigars. Thomson says that others

> looked up to him on account of his literary attainment; he was useful as a master of ceremonies in the presence of native officials, and he kept a careful journal. He esteemed himself our protector, and it was truly gratifying to notice how he courted the society of officials to whom we had credentials, and before whom our interpreter exposed us, at the same time introducing to their notice our foreign wine and our cigars – commodities with which he had been laudably sedulous to make himself acquainted beforehand (Thomson 1874).

Thomson's presentation of his photo of Chang is as follows:

> There he is, presented to the reader in No. 35, just after he had been droning, in an obscure corner of the cabin, over a whole classical commentary. The figure to the left is one of the boatmen, while a Ningpo boy is looking from the cabin door; the characters are faithfully rendered, and are engaged in the several occupations with which half their time was thickened. The boat was under weigh-in when I executed this picture (Thomson 1874).

Chang liked to drink. On the first night of Chinese New Year, when all the crew were celebrating, Chang protested that "his honourable name had been sullied by the drunken behaviour of the boatmen; I however, discovered quickly that our venerated interpreter was himself not without sin, being, indeed, unable to stand erect" (Thomson 1874).

However, Chang had his uses. When the peasants, unaccustomed to seeing outsiders and suspicious of photographic equipment, throw stones and other objects at the foreigners and Thomson narrowly escapes being struck by an oar, Chang tries in vain to reason with the crowd.

However, Chang is not always so brave, and, contrary to what the image might suggest, he is something of a landlubber. Descending the rapids at Shan-tow-pien, the other boatmen convince Chang to join them, but "as the vessel plunged and groaned in an agony of straining timbers, he became perfectly sick with panic fear" (Thomson 1874).

So many of my ideas about Chang were wrong. First, he was not the boat's pilot or guide. Thomson employs Captain Wang as a pilot, and Chang's duties seem limited to those of a protector of sorts, especially since he is unable to communicate with the populations on the riverbanks. In the photo we can see Chang's sense of superiority. That he took advantage of the wine and cigars of Thomson and the Americans doesn't surprise us, but it is more difficult to imagine his love of literary classics. The photo was taken shortly after Chang had "recited and read his literary classics". Perhaps he was reciting a Buddhist sutra as this would explain the *japamala* he is holding.

After appearing so confidently at the stern of the boat, as if he was commanding everyone, the biggest surprise is his fear of the rapids,

The interpretation of art or photography is always individual, and we can question the idea of discovering the "essence" of the person being photographed. If we return to the observations about photography in the Introduction, can we not see a contradiction? Can a photograph "bring out the essential characteristics of any person presented [with] dignity and depth of their perception" (Kracauer 1980)? Can it show the "essence" of a person: "Most of my photos are grounded in people, I look for the unguarded moment, the essential soul peeking out, experience etched on a person's face" (McCurry 2020). Or, on the other hand, will it reflect "the infinite variety of subject matter offered by the natural universe" (Brassai in Hill and Cooper 1992), and develop a "complex intertextuality" (Burgin in Clarke 1997:27), a range/variety of multiple and indistinct meanings" (Kracauer 1980), and "phenomenological doubt", that is, "several different points of view" (Flusser 2000)? The answers to the questions on Chang points us to the wide range of interpretations of a photograph.

My impressions of Chang were quite different from those of many of the survey respondents, and we can return to Barthes again, but now he is admitting the ambiguity of photography: "Photography never lies: or rather, it can lie as to the meaning of the thing, being by nature tendentious, never as to its existence" (Barthes 1981:87).

5. More on John Thomson

John Thomson was Official Instructor in Photography for the Royal Geographic Society from 1886, which he had joined in 1886, and called for a 'geographical photography'. In a lecture on photography and exploration to the British Association for the Advancement of Science in 1891 he claimed:

> Where truth and all that is abiding are concerned, photography is absolutely trustworthy, and the work now being done is a forecast of a future of great usefulness in every branch of science. What would one not give to have photographs of the Pharaohs or the Caesars, of the travellers, and their observations, who supplied Ptolemy with his early record of the world, of Marco Polo, and the places and people he visited on his arduous journey? We are now making history and the sun picture supplies the means of passing down a record of what we are, and what we have achieved in this nineteenth century of our progress (in Ryan 1997:24).

He saw himself very much as a scientific explorer interested in the scientific application of photography (Ryan 1997:25). The great majority of his photos attempted to define racial types, features of gender, age social classes, and occupation, and also landscapes, "a complete and objective picture of China's landscape and people" (Ryan 1997:62-63), for example, "Four Heads, Types of the Labouring Class", plate XI, volume 1" (Hockley:Internet). His photographs along the Yangtze were made with the specific design of showing the commercial possibilities of opening up the river for (British) shipping (Ryan 1997:65) after the Second Opium War (1856-60), "bringing an inevitable extension of the 'light' of Western civilization, through commerce and Christianity, into the 'darkness' of the East" (Ryan 1997:63), at a time when Britain was consolidating its Empire, on which the sun never set.

This attempt to use photography to portray certain "types" can be seen in the complementary photo to "Chang", that of the boatmen (*Photograph 2*). Thomson comments that they were lucky to escape the noxious odours of Chinese cuisine and that the boatmen never changed their clothes, smoked tobacco or opium to sleep, and never took off their padded coats except in the morning to try to get rid of some of the many fleas (Thomson 1874).

By contrast, Chang cannot be categorized as a type: is he an interpreter, a guide, a charlatan, a scholar, an actor? Likewise, Thomson does not seem to be trying to pigeon-hole him but rather open up to his individuality and allow him to be interpreted in a variety of ways.

Photograph 2: The Boatmen

4. The Well-Dressed Young Indigenous Interpreter

Photograph 3: The Young Well-dressed Indigenous Interpreter

We have much less information about the next photograph, *The Young Well-Dressed Young Indigenous Interpreter*. I found the photo on the social network Reddit together with the following caption in Portuguese: "Caripuna Indian interpreter of Cândido Rondon during expeditions in Amapá. Photograph from 1914" ["Índio Caripuna intérprete de Cândido Rondon durante expedições no Amapá. Fotografia de 1914"]. The photo was posted in 2017 and is followed by a small number of comments on his good looks, elegant clothes and how hot he must have felt.

A little research disproved this information. i) Amapá only began to exist as a territory of Brazil in 1943. However, the name Amapá had been used before to refer to the region that is today the state of Amapá; ii) In 1914 Marechal Rondon was finishing his expedition with former president Theodore Roosevelt and later worked on installing the Mato Grosso-Amazonas telegraphs. He only went to the border region between Amapá and the Guianas in 1927-1928 but was in Manaus in 1914.

The photograph is excellent quality, and it seems that the young man does not feel uncomfortable in his suit and looks directly at the camera, giving the impression that he is enjoying being photographed. In *Memória do SPI: textos, imagens e documentos sobre o Serviço de Proteção aos Índios (1910-1967)* (Freire 2011) we see the photos of Rondon and the SPI posts, and we see all the indigenous people in the posts wearing Western clothes, hiding their nakedness, the men in long trousers or shorts, and the women in long Mother Hubbards to hide the shape of their bodies. But there is no photograph that shows an indigenous person dressed as elegantly as this young man. In fact, the confidence he shows facing the camera gives us the impression that he is used to being photographed and wearing suits, that this experience is nothing new, and that he is at home in the urban environment.

We wonder where the photo was taken. It was certainly in a good quality professional studio, which gave the photo an excellent finish. It is doubtful whether there were studios in Amapá at that time, so did they have to go to the regional capital of Belém? And whose suit is it? Was it given to the young man as payment for his interpreting services? Or did he already have his own suit to wear on Sundays? Or it belonged to the studio, which lent it to those being photographed.

The haircut is modern, and if the photo didn't have a studio background, and the young man were wearing other clothes we might think the photo was contemporary.

Unlike Vegnon in *Photograph 4*, Chapter 5, who seems to be uncomfortable with his outsize uniform from the Paraguayan War, from some forty years ago, the suit seems to fit him well and to have been tailored; or almost. Maybe the sleeves are a little long, but it's hard to know exactly. He gives the impression that he is very proud to wear his suit and enter the Western world.

4.1. The Questionnaire

A similar questionnaire to that given on the photograph of Chang was applied to the same postgraduate students of Translation Studies.

1. Where do you think the photograph was taken?

2. When do you think the photo was taken?

3. What might this interpreter's work have been like?

4. Does the interpreter give the impression of being a good interpreter? Why?

5. In your own words, describe the interpreter.

6. What other elements of the photograph are of interest?[33]

The majority had a good impression of him, describing him as an important and self-confident professional, with good education, an observant look, serious and responsible, already accustomed to the job and dealing with white Europeans, while a minority had a more negative opinion.

Others failed did not show any feelings towards the interpreter, describing him merely by his clothes. And finally there were those who responded that he was not the interpreter as the professional interpreter does not appear in the photo

More comments were made in the last question on the interpreter's elegant suit, but others mentioned the suffering look that the interpreter seems to have, with a forced unnatural smile, as if he were. The background was also

---

[33] "1. Onde você acha que a fotografia foi tirada? / 2. Quando você acha que a foto foi tirada? / 3. Como podia ter sido o trabalho deste intérprete? / 4. O intérprete dá a impressão de ser bom intérprete? Por quê? / 5. Em suas palavras, descreva o intérprete. /6. Quais outros elementos da fotografia são de interesse?"

mentioned, but several respondents did not realize it was a false background in a photography studio.

The discussion above demonstrates, as in the case of Chang, the difficulty of reaching a consensus about a photograph, and the wide variety of impressions that a photograph can give. Is the young person happy or embarrassed about being photographed? Does he feel good wearing the suit? Is he proud to have received the suit as part of his payment for performing his role as an interpreter? Or is this a set up?

4.2. "Civilized" Indian

I tried to get information from several experts in the field. I discovered that the photo was used on a panel, "*Índios 'Civilizados' no Rio de Janeiro*", [Civilized Indians in Rio de Janeiro], part of the exhibition *O Rio de Janeiro continua índio* [Rio de Janeiro remains Indian] organized by the Museu do Índio/FUNAI and the Museu da Justiça do Estado do Rio de Janeiro in August 2015, under the curatorship of Carlos Augusto da Rocha Freire, which afterwards moved to the Universidade do Estado do Rio de Janeiro in April 2016. I obtained the following information from curator Ana: "I looked in the catalogue of the exhibition we held, and it says the following about the photograph: Photos from the Rondon Commission, s/d. The photo is in the part of the 'Civilized' Indians in Rio de Janeiro, and exemplifies cases of Indians brought by Rondon from Mato Grosso to study and work here in Rio"[34] (in José Ribamar Bessa Freire 2020, e-mail).

On 29 November 2023 I tried some crowdsourcing, sending the following message together with *Photograph 3* to a number of translator e-mail lists:

> The photo is part of the study I am carrying out on photographs of interpreters, which can be seen in the Lexikos book:
>
> Fotografias de intérpretes: em busca de vidas perdidas
> https://lexikos.com.br/produto/fotografias_de_interpretes/?v=9a5a5f39f4c7
>
> I am now preparing the version to be published in the UK.

---

[34] "Olhei no catálogo da exposição que fizemos e consta o seguinte sobre a fotografia: *Fotos da Comissão Rondon, s/d*. A foto está na parte dos Índios 'Civilizados' no Rio de Janeiro, e exemplifica casos de índios trazidos por Rondon do Mato Grosso para estudar e trabalhar aqui no Rio".

I obtained very little information about the attached photo. Does anyone know the photo?

Reddit says he is an interpreter of Mal. Rondon in Amapá in 1914, but Rondon was not in Amapá in 1914. The photo was used in a panel "'Civilized' Indians in Rio de Janeiro, part of the exhibition "Rio de Janeiro Remains Indian" by the Museu do Índio/FUNAI and the Museu da Justiça do Estado do Rio de Janeiro in 2015. The curator Ana wrote to me that he was probably from Mato Grosso and taken to Rio to study and work.

Thanks in advance,

John[35]

The most important reply was from the professional translator Renata Schinke, who had found a photograph of Rondon with someone who looked very much like the young man and a group of Indians. I discovered this photo was published in the *Jornal do Comércio* in 1915, with a caption that told us he was visiting the Paresí in Mato Grosso in 1915 (*Photograph 4:The Paresí are happy to see Rondon with them*). The young man here is also wearing a white jacket, which may be that of the studio photograph, *Photograph 3*. So, our findings help us conclude that indeed he is an interpreter but from Mato Grosso rather than Amapá and speaks Paresí and therefore almost certainly a member of the Paresí tribe. I also included *Photograph 5: Presents of beads are distributed to the Indians*, similar to *Photograph 4*, but it has not been cropped and gives a clearer view of the young man.

---

[35] Informação sobre a fotografia em anexo
A foto faz parte do estudo que estou fazendo sobre fotografias de intérpretes, que pode ser visto no livro da Lexikos
Fotografias de intérpretes: em busca de vidas perdidas
https://lexikos.com.br/produto/fotografias_de_interpretes/?v=9a5a5f39f4c7
Agora estou preparando a versão que vai ser publicada no Reino Unido.
Consegui muito pouca informação sobre a foto em anexo. Será que alguém conhece a foto?
Reddit diz que ele é um intérprete do Mal. Rondon em Amapá em 1914, mas Rondon não esteve em Amapá em 1914. A foto foi usada num painel Índios 'Civilizados' no Rio de Janeiro", parte da exposição O Rio de Janeiro continua índio do Museu do Índio/FUNAI e o Museu da Justiça do Estado do Rio de Janeiro em 2015. A curadora Ana me escreveu que ele provavelmente era de Mato Grosso e levado ao Rio para estudar e trabalhar.
Thanks in advance,
John

Photograph 4: The Paresí are happy to see Rondon with them

Photograph 5: Presents of beads are distributed to the Indians

CHAPTER 6

JOHN BROWN:
THE INTERPRETER IN THE DEVIL'S PARADISE

Unlike the sections on Chang and The Well-Dressed Young Interpreter in the Chapter 5, in this chapter we have a good deal of, if sometimes contradictory, information about our interpreter, John Brown (1879-1977), interpreter on Captain Thomas Whiffen's 1908 attempt to discover what happened to the French explorer Eugène Robuchon (1872-1906), who had mysteriously disappeared a few years earlier. Brown was also the interpreter between English and indigenous languages for foreign consuls George B. Michell, British Consul in Peru, and Stuart J. Fuller, from the United States, who visited the Putumayo region of the Peruvian Amazon on the border of Colombia with Brazil, from October to December 1912, after the publication of the report by Sir Roger Casement (1864-1916) on the atrocities – the mistreatment and murder of indigenous people working in rubber extraction – of which John Brown was one of the main witnesses. Casement's report had been sent to London and was published in July 1912 as *The Blue Book*. Indeed, in July 1911 he was knighted for his humanitarian work.

1. John Brown: one (of various) biographies

There are several sources about Brown's life. Wylie (2013), and Lino e Silva and Wardle (2016) tell the biography of John Brown, following the Colombian Pedro Gómez Valderrama (1984). However, the most reliable basis should be the article by Brown's grandson, Ramiro Rojas Brown, "John Brown: un personaje de leyenda y testigo de excepción" ["John Brown: a legendary character and exceptional witness"] (2014), which I use as the basis of the following pages.

Brown was born in Chicago on 23 September 1879, the grandson of southern slaves, the fourth child of Elizabeth and William Brown, and his family had been part of the great northward migration following Emancipation. He grew up near the port city of Chicago, in an area much visited by foreign sailors.

He was almost certainly named after his famous white namesake, the militant anti-slavery activist, who had died twenty years earlier. John Brown's father died when the boy was seven, and, at the age of ten, Brown sailed on a ship to New York, where, helped by sailors, he sailed to São Tomé, Barbados, Trinidad, and then Guadeloupe, and finally Le Havre, France, where he survived by washing dishes in hotels, and then Paris, where he washed dishes at Le Grand Chavert and Le Grand Marché hotels.

Then he travelled to Liverpool, where he obtained a sailor's licence, met another black sailor, James Henry, and his white girlfriend, Margaret, who taught John to read and write so he could send letters to his mother in Chicago.

Now he was fifteen, and he went to sea again, this time as a stoker's mate, working in the engine room on ships for the White Star Line, the Royal Mail, and other companies that plied between the islands of the Caribbean, Africa, and the UK for several years before finally becoming a stoker in his own right.

While on vacation in Glasgow, a sailor encouraged Brown to escape the dirt and heat of the engine room and travel to the Amazon to work in the booming rubber industry. So in 1902 Brown took a boat from Manaus to Iquitos, Peru. There, in a bar, he saw an Amerindian woman for the first time and was fascinated. His companions told him that there were many like her living naked in the forest, and, to see them, he entered the Putumayo jungle and started working at Júlio César Arana's Casa de Arana – the Peruvian Amazon Company (PAC) – as a rubber tapper, despite being warned by a Barbadian worker: "Bad. They kill people there. There are wild indians. They say people never come back" (Gómez Valderrama 1984, in Lino e Silva and Wardle 2016).

From 1902 to 1905 he worked at PAC, in the Abisinia section, and his work consisted of taking part in the "comisiones" to hunt new Indians and take them to the section to collect rubber. In the two years and three months he spent in Abisinia, he saw hundreds of dead Indians, and the Indians' territory always being invaded.

At Casa Arana's other rubber extraction sites, he saw the same atrocities and began to understand Casa Arana's philosophy:

> For the bosses of the rubber company, the lives of its employees were of no value, especially if there were suspicions that they could undermine the name and aims of the company, which for Brown was to obtain maximum

profit at the expense of the slave labour of the indigenous people. and the appropriation of disputed territory, with the complacency and complicity of some Colombian politicians, to whom, according to Brown, the Aranas sent money monthly[36] (Rojas Brown 2014).

In 1906 John Brown was a member of the group that tried to find the French explorer Eugène Robuchon, who had been lost in the jungle and possibly murdered by the PAC. Brown said that Robuchon had taken pictures and intended to write a book, which could incriminate Arana. Brown assumed that the search expedition was just a decoy to cover up the fact that he had been murdered (Rojas Brown 2014).

In 1907 he began working in the La Chorrera section as a carpenter and later as a messenger in another section of the PAC in El Encanto, when he met the American engineer Walter Hardenburg (1886-1942). Brown and other rubber workers handed him papers that told of the suffering they and the natives had been through. Indeed, Hardenburg was the first to denounce the PAC in the British magazine *Truth* in 1909, the consequence of which was the visit and subsequent report by Roger Casement (Rojas Brown 2014).

In June 1908 Brown received permission to go to Iquitos to send money to his mother; there he sought out the British consul, David Cazes, telling him of the atrocities, but Cazes had extensive contact with Arana, and when he subsequently went to London, he made no denunciations.

2. Whiffen's "Boy"

Brown met Captain Thomas Whiffen in Iquitos in 1908, when, on his own account, Whiffen intended to leave Putumayo for good and return to the United Kingdom. Whiffen's aim was to find Eugène Robuchon and study the "cannibal tribes" of the northwest Amazon, published in 1915 in *The North-West Amazons: Notes of some months spent among cannibal tribes*. But they failed to find Robuchon. In contrast, Brown's intention in becoming Whiffen's helper, as he explained to Gómez Valderrama many years later, was for Whiffen to witness the crimes being committed by the

---

[36] "Para los jefes de la empresa cauchera la vida de sus empleados no tenía ningún valor, y pior si se intuía que podían atentar contra el nombre y los propósitos de la empresa, que para Brown, era lograr el máximo lucro a expensas del trabajo esclavizado de los indígenas y apropriarse de un território en litígio, con la complacencia y complicidad de algunos políticos colombianos, a los cuales, según Brown, los Aranas mensualmente mandaban dinero".

House of Arana and to take news of the atrocities to London (Gómez Valderrama 1984, in Lino e Silva and Huon Wardle 2017).

Whiffen describes his relationship with Brown with a certain condescension, always calling him "My boy".

> I had with me at that time John Brown, a Barbadian negro. He had been in the district of Issa for about three years, employed by a Rubber Company, and I enlisted him as my personal servant in Iquitos. He had "married" a Witoto woman about two years before, and through that connection I was able to gain a lot of valuable information. Indeed, he was invaluable throughout the expedition and was more loyal and devoted than a traveller who had any experience of dealing with African boys in their own territories had reason to believe any Negro servant (Whiffen 1915:3).

John Brown says Whiffen spoke to him in Iquitos and said he wanted to find out how Robuchon had lost his life. Brown replied that he would accompany him on the condition that he himself be the leader: "If you let me be your boss": "I'll tell him where to go, what to do [...] If they [the Casa Arana foremen] found out that we were going on an expedition looking for Robuchon, I would be dead and so would he" (from an interview with Jorge Gasché and Mireille Guyot in 1969 in Puerto Leguízamo, in Echeverri 2019).

In this way Brown would decide which places to visit so as not to arouse suspicion. Brown spoke the Murui language and knew a little of the Bora language, in addition to Spanish and English. Whiffen's Spanish was basic, and he had no knowledge of native languages. In addition to the quote above, Whiffen is grateful for Brown's knowledge of languages, which "were of great use to me as a basis upon which to work" (Whiffen 1915:249).

Brown's knowledge and experience in the region were decisive for Whiffen to collect information in those seven months on horticulture, fishing, hunting, social organization, rituals, diseases and cures, funerary rites, shamanism, dancing and singing, and religion and beliefs – which make up a large part of his book – and which were obtained through the intermediation of Brown, perhaps the real ethnographer of this work. His dependence on Brown can be traced through a nightmare Whiffen had of being lost in the forest for some hours without Brown's company: "Then, after two fruitless hours of growing despondency, turned and went down, to find, as darkness was closing in, Brown and his party. That night I had fever, and talked in my sleep. And John Brown was lost for five and a half months. Good God!" (Whiffen 1915:38)

Whiffen had made some references to the problems in the *Gazette of the Royal Geographical Society* and in *The Daily News*, and Brown had hoped that Whiffen's book would communicate the details of Indian cruelty and exploitation, but Whiffen failed to mention these problems, and his text describes pure indigenous peoples who live isolated from civilization, using photographs to validate this image (Echeverri 2014).

3. The Devil's Paradise

Brown visited his family in Chicago in early 1910 and was later in the Caribbean, where he made a statement about mistreatment to the governor of Barbados. While there he received a message from his mother saying that Roger Casement wanted to hire him as an interpreter; however, Brown arrived too late in the Putumayo in early September 1910, and Casement had already hired the Barbadian Frederick Bishop. Nevertheless, Brown made an important statement that was included in Casement's report to the British government, *Correspondence Respecting the Treatment of British Colonial Subjects and Native Indians Employed in the Collection of Rubber in the Putumayo District* (1912). Casement considered the enslavement and massacre of Indians in the Amazon region of Peru "a crime against humanity", perhaps using this expression for the first time ever (in Lino e Silva and Wardle 2016).

Casement was initially distrustful of Brown. Brown was not from the UK but apparently from Barbados, a British colony in the Caribbean, and he was black. He seemingly had British citizenship as he was from a British colony. Was it possible to trust him? Did his statement have much value?

> This account of Whiffen's "boy" was promised us some time ago, but now that it has arrived it is difficult to say how much credence should be accorded to his statements (especially as the exceptionally good English he writes is quite suspect, in the case of blacks from West Indies) [...]

> Can we send a copy of John Brown's statement to the company [Peruvian Amazon] saying that, in our opinion, this strengthens the case for a commission of inquiry [...]

> [The Peruvian Amazon Company] would merely reply that Captain Whiffen's black servant was not a reliable witness [...] (Peruvian Amazon Co.1910, in Lino e Silva and Wardle 2016).

Casement had an initial negative impression of Brown:

> John Brown demands thirty pounds that the Company owes him and claims
> that over a hundred Barbadian workers continue to work in inaccessible
> places, "kept like slaves". Probably, almost certainly, a lie, but there may
> still be some of them in remote outposts like Morelia or Abisinia[37] (in Izarra
> and Bolfarine 2016).

Brown's testimony was one of the most important for Casement to bring to
light the atrocities of the PAC and confirm the presence of Barbadians
employed by the PAC in Putumayo. Brown had worked with the
"muchachos", usually "trustworthy Indians", but also Barbadians,
responsible for the discipline and punishment of recalcitrant Indians, under
the command of Sr Agüero. Brown participated in many "tasks", the first of
which was in 1905, when they went to capture Bora Indians to work in the
rubber plantations. The Bora were wild and ferocious, and many of them
had not been conquered, according to Frederick Bishop, Casement's
interpreter, and other Barbadian witnesses. Brown's party captured six
women, three men and three children, and killed six other Indians: a small
boy, shot in the stomach as he tried to flee, the chief, whom they shot, and
three men and women, whom they decapitated, holding them by the hair
while their heads were chopped off with machetes.

They took the prisoners to the *cepo*, the stump, for punishment. Later one
of them was shot and killed by Agüero. The rest fled. One of them was under
Brown's guard and was carrying rice from the port on a two-day trek.
During his stay in the Abisinia rubber plantation for two years and three
months, Brown was mostly tasked with chasing down escaped Indians and
saw hundreds of them being killed (Taussig 1993). In Brown's words:

> [...] they were decapitated, dismembered, tortured by flogging them in the
> trunk, where they died of hunger and thirst; sometimes they were burned
> alive and sometimes tied with their hands behind their backs and thrown
> into the river. He saw women beheaded nursing their children, and children

---

[37] John Brown exige trinta libras que a Companhia lhe deve e afirma que mais de
cem trabalhadores de Barbados continuam trabalhando em lugares inacessíveis,
"mantidos como escravos". Provavelmente, quase com certeza, mentira, mas pode
haver ainda alguns deles em postos remotos como Morelia ou Abisinia (in Izarra e
Bolfarine 2016:49).

murdered and hacked to pieces; this, in a "comisión" commanded by Esteban Angulo, who carried out the crimes[38] (in Rojas Brown 2014).

In addition, Brown saw many Indians who starved to death (United States Congressional Serial Set 1913). He left Abisinia in October 1907 and returned to La Chorrera, where he spent the rest of his time in the PAC until he went to Iquitos in June 1908, when he met Captain Whiffen. At La Chorrera he witnessed flagellations, which had apparently been hidden from Whiffen (United States Congressional Serial Set 1913).   •

Furthermore, the Barbadians themselves had been tortured. As he wrote in the letter he sent to the Colonial Office in March 1910:

> They beat us with swords, they put us in guns (hands tied across knees with guns underneath knees), and did us all manner of wickedness. We cried for help but there was none. We tried to escape, but there was no means of doing so – only one small steamer which belonged to the same company and they would not take us away. They still continued and beating us English subjects, and they treated the Indians in the same way. They (the Spaniards) would take a man, tie his hands together behind his back with chains and hang him up, and then beat him with sticks or swords (Goodman 2009).

4. John Brown, Interpreter of the Consuls

Nevertheless, John Brown's most important interpreting assignment was as in interpreter between English and indigenous languages for foreign consuls George B. Michell, British Consul in Iquitos, and Stuart J. Fuller, US Consul in Iquitos, who visited Putumayo from 7 August to 6 October 1912 in order to check the situation and see if conditions had improved after Casement's report was sent to London and published as an official report in *The Blue Book* in July 1912 (Mitchell 2003). However, the consuls were accompanied by Carlos Rey de Castro (1866-?), Peruvian consul in Manaus and friend of Arana, Pablo Zumaeta, Arana's employee and brother-in-law, the Portuguese photographer Silvino Santos (1886-1970), and Arana himself (1864-1952) during the three months, which would make his work very difficult (Gray 2005). During that time John Brown was the guide and

---

[38] "[…] decapitaba, descuartizava, se torturaba mediante flagelaciones en el cepo, donde morrían de hambre y de sed; algunas vezes eram queimados vivos, y otras, amarrados de las manos en la espalda y tirados al río. Vio a mujeres decapitadas amamantando a sus hijos y niños asesinados cortados en pedazos; esto, en una comisión comandada por Esteban Angulo, quien realizó los crímenes".

interpreter, and, according to Ramiro Rojas Brown, "At no point did Arana try to intimidate him"[39] (Rojas Brown 2014).

We see the results of the consuls' visit in the reports included in *La defensa de los caucheros* [The defence of the rubber farmers] (Rey de Castro et al. 2005). At the end there is a statement by John Brown, who says that Michell did not investigate any of the events that had taken place years before and did not renew Casement's investigations. If anything had happened, it was in the past, and the group's plan was to investigate the current state of things (Rey de Castro et al. 2005).

In his statement Brown insists that they tried to see the true state of things and the treatment that the natives received. They had contact with many indigenous people, and Brown translated the questions and answers for Michell and Fuller. Brown gives an example of a typical conversation:

> Q: Why do you work with rubber?
> A: Because we want shotguns, clothes, axes, machetes.
> Q: How much rubber do you need to bring to pay for what you get?
> A: For rifles we bring more than for trousers, for trousers we bring this much (making the gesture of a small amount).
> Q: If you don't bring rubber, are you mistreated, punished or beaten for that reason?
> A: No. They don't mistreat us, even when we don't take anything.
> Q: What do you have to eat?
> A: Cassava, curassow, paujil birds, and piuri, banana etc.
> Q: Do you have enough to eat?
> A: Oh, very much.[40]
> (Rey de Castro et al. 2005).

The conclusions are that "The Indians are happy, they work effortlessly, they are well fed and, every day they improve their primitive conditions of indolence and life in the jungle, with agricultural work" (Rey de Castro et al. 2005). And John Brown signs noting that he attended all the investigations

---

[39] "En algún momento Arana trató de intimidarlo".

[40] P Por qué es que ustedes trabajan caucho? / R Porque queremos escopetas, vestidos, hachas, machetes. / P Qu. cantidad de caucho necesitan ustedes traer para pagar lo que les dan? / R Por escopeta traemos más que por pantalón, por pantalón traemos as (significando com las manos un manojo de pequeñas dimensiones). / P Si no traen caucho son ustedes por ese motivo maltratados, castigados o flagelados? / R No. No nos maltratan aun cuando no traigamos nada. / P Que tienen ustedes para comer. / R Casave, paujil, piuri, plátanos etc / P Tienen ustedes bastante que comer? / R Oh, mucho.

as the only interpreter and that the report is a faithful translation into Spanish of what he said in English.

However, Ramiro Rojas Brown tells us that "Despite all the effort made to hide the reality of the events that occurred, the commission found more negative elements that increased the number of complaints presented"[41] (Rojas Brown 2014).

It is strange that, a few months after making his statement to Casement about the atrocities that took place in Putumayo, he signed the declaration about the happiness and good life of the Indians employed by Arana. Surely the money he was going to receive had some influence, but doesn't his statement give the impression that the problems have already been resolved? Wouldn't Brown be a pretty obvious target for Arana's henchmen? Well, after that visit, Brown, already in Iquitos, no longer an employee of the Arana company, received the news that there was a murderer after him. He went to Brazil and then to Colombia, where he lived for the rest of his life, working in rubber and balata latex extraction. In 1934 he fought on the side of Colombia in a border war with Peru and, as a result, was given Colombian citizenship (Rojas Brown 2014).

In 1971, looking for hallucinogens, the North American researcher Terence McKenna met John Brown at his house, next to the Putumayo River, in Colombia:

> We soon reached a ramshackle and undistinguished house with a small yard hidden behind a tall board fence… A large pig lay in the lowest, wettest part of the yard; three steps up was a veranda. Upon the veranda, smiling and motioning me forward, sat a very thin, very old, much wrinkled black man: John Brown. It is not often that one meets a living legend and, had I known more about the person I confronted, I would have been more respectful and more amazed. "Yes", he said, "I am an American". And, "Yes, hell yes, I am old, ninety-three years. Me hee-story, baby, is so long". He laughed dryly, like the rustle of roof thatch when tarantulas stir (McKenna 1993) (in Moises Lino e Silva and Huon Wardle).

---

[41] "A pesar de todo el esfuerzo realizado por esconder la realidad de los hechos ocurridos, la comisión encontró más elementos negativos que aumentaron el cumulo de denuncias realizadas".

5. Brown Chameleon

Brown was born in Chicago, but told everyone he was Barbadian, or that he came from the tiny island of Montserrat, so that he could receive protection as a British citizen.

In his 1910 statement to the Commissioner of Montserrat, Brown said that he had arrived in Iquitos in 1902. Later in the same year in Iquitos, however, he would tell Roger Casement that he had arrived in Putumayo with a group of "Barbadian" companions hired in Bridgetown in 1904 and 1905 (Johnson 2002 in Lino e Silva and Wardle 2016). Probably Brown lied for a specific reason: his story had to be similar to the stories of other black Caribbean workers in Putumayo, to protect the rights of "British subjects" (Lino e Silva and Wardle 2016).

For Lino e Silva and Wardle (2016) Brown, this chameleonic characteristic was typical of many other Caribbeans:

> […] black migrants like John Brown, by way of many decades of travel across the Caribbean and Central and South America, had become experts at a form of cultural mimesis, remodeling and code-switching that allowed them to reshape the public expression of their worldviews and motives to fit with the signs of imperial power (Lino e Silva and Wardle 2016).

This change, this self-remodelling, can be seen in the way in which Brown, a few months after the publication of Casement's report in July 1912, in which his statement of the Putumayo atrocities was of paramount importance, worked as interpreter and guide for the inspection of the consuls of Peru, the United Kingdom, and the United States, accompanied by Arana and, at the same time, his statement was very positive regarding the treatment of the indigenous people.

Gómez Valderrama retells Brown's life story as a series of personal escapes from one 'Hell' to another (1984). I prefer to see John Brown as a kind of symbol of cleverness, a trickster, managing to communicate to the world the atrocities he witnessed and in which he participated, escaping a threatening situation, enjoying a life of almost a hundred years and, according to one source, being the father of 18 children (MiPutomayo 2013).

6. The Photographs of John Brown

Photograph 1: John Brown with Parrot (1912)

Photograph 2: John Brown, "Whiffen's Boy" (1908)

Photograph 3a and 3b: Whiffen and Brown Dressed as Indians (1908) and (1912)

Photograph 4: John Brown in Top Hat (1908)

*Photograph 5: John Brown with Daughter and the Anthropologist
Horacio Calle Restrepo (1968)*

Photograph 6. John Brown on Book Cover (1964)

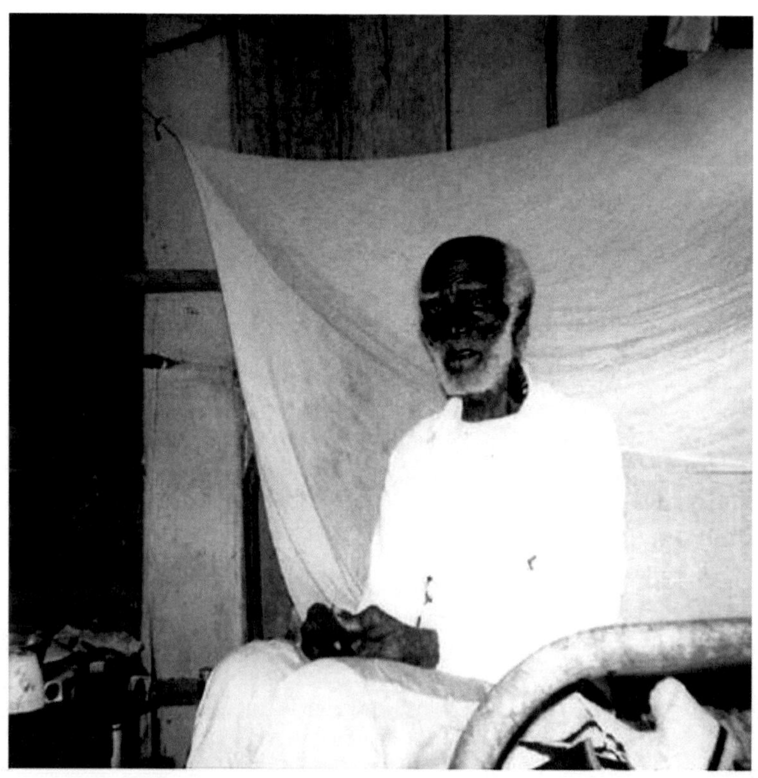

Photograph 7. John Brown, Very Old (c.1975)

Photograph 1: John Brown with Parrot, and *Photograph* 3b, showing John Brown dressed as an Indian, come from *Album de Fotografías: Viaje del Comisión al Rio Putumayo y sus Afluentes, Agosto a Octubre de 1912* [Album of Photographs: Trip of the Commission to the Putumayo River and its Tributaries, August to October 1912] (Chapparro and Chirif 2013), available on the Internet, This album is a selection of photographs commissioned by Arana from the Portuguese photographer based in Brazil, Silvino Santos, to show peace and harmony and the "civilizing role" (Chapparro and Chirif 2013) of the company. But, for Chapparro and Chirif, the photos have "the intention of distorting, manipulating these subjects"[42] (Chapparro and Chirif 2013), and they mention Jean Pierre Chaumeil's "war of images" (2009), which "devalues the figure of the subaltern subject and

---

[42] "[…] la intención de distorsionar, manipular a estes sujetos".

idealizes that of the rubber boss, showing them in an atrocious and fragile harmony: they all pose before the lens of the Portuguese photographer"[43] (Chapparro and Chirif 2013).

We see photos of Indians preparing food, breakfast and lunch (Chapparro and Chirif 2013:74) and organizing parties and dances (Chapparro and Chirif 2013). There are pictures of women in long dresses, the wives of the "muchachos", both the huitotas (Chapparro and Chirif 2013) and the boras (Chapparro and Chirif 2013), one shows a woman working with a Singer sewing machine (Chapparro and Chirif 2013), and another shows boys and men working on the rubber pressing machine (Chapparro and Chirif 2013). The photos bear certain similarities to the official photos from the Indian Protection Service (SPI), which I referred to in Chapter 3. There are photos of the managers with the Indians, all in apparent harmony (Chapparro and Chirif 2013). The Indians are thin, but they look strong. There are no old or sick Indians. They even look healthier and happier than the white officials and consuls, who have to work in the Amazonian climate in a suit and a high stiff collar (Chapparro and Chirif 2013).

In Chapters 1 and 2 I emphasized the way in which the interpreter brings people together, the French and the Chinese, and the Americans with the native Indian tribes, although with certain caveats in the second case: did the North American Indians really want to get into the capitalist system? But here, the "space" in which John Brown acts, dominated by the financial interests of the Peruvian ruling classes and international capital, especially British, resulting in the atrocious treatment of the indigenous people of the region, appears to have more similarities with the "contact zone" that Mary Louise Pratt describes: "the space of imperial encounters, the space in which peoples geographically and historically separated come into contact with each other and establish ongoing relations, usually involving conditions of coercion, radical inequality and intractable conflict" (Pratt 1992:4; also in Langford 2014).

Here we can place John Brown in the "contact zone" space and compare him to the interpreters in Langford's article, "Framing and Masking" (2014), described in Chapter 1. And there are certain similarities with the pictures in Langford's article that show the isolation and alienation of the interpreter. *Photograph 1: John Brown with Parrot,* shows him alone, just with his

---

[43] "[…] desvaloriza la figura del sujeto subalterno e idealiza la del patrón cauchero, mostrándolos en una atroz y endeble armonía: todos posan ante el lente del fotografo portugués".

parrot for company, apparently without comrades although a Barbadian appears in the background, looking at him, but not keeping him company. Brown occupied the middle tier, between the bosses and the Indians, in very precarious conditions and probably disliked by both groups. In the album we see the photos of Arana, Michell, Fuller, Carlos Rey de Castro, eating and drinking with the captain of the boat, Ubaldo Lores (Chapparro and Chirif 2013) on one hand, and, on the other hand, the Indians, in the photos apparently happy. Brown belonged to the middle group, the "muchachos", the executioners and butchers. The indigenous people gave the Barbadians a name, *taife*, which is close to the meaning of "devil" (Pineda, 2000 in Cabrera Becerra 2018).

As in many of the cases we examine throughout the book, the interpreter comes from a lower class, and in this particular case, the interpreter Brown is black; the work is completely oral, there is never a need to read or write the languages he is working with; and the tasks of interpreting are combined with that of scout – for Whiffen Brown had to interpret both the indigenous languages and the terrain, the jungle and dangers along the way – both the flora and fauna. Casement shows us this class division in the separation of tables on the boat on which he traveled:

> I discover that Garrido, Gusmán and the other employee are eating at the table with us! At the posts, these "turkeys" are forced to eat separately and, when we went up the river, Garrido ate with Bishop. Bishop and John Brown sit at a side table on the upper deck, where servants once ate their meals on the voyage upriver. Finally, Arédomi and Omarino are on the upper deck, as is a small chiviclis [small rodent] that Macedo got for me.(Casement 2016).

In *Photograph 2: John Brown, "Whiffen's Boy"*, we see Whiffen and Brown on their expedition. Brown is respectful, a step behind his boss, and his clothes are is even more ragged than those of Whiffen's. The photo series demonstrates this chameleonic side of Brown. Here he seems to be the dutiful servant, "Whiffen's Boy", although, as we have seen, he was the one who made all the important decisions regarding the expedition and provided much of the information on the flora and fauna for Whiffen. And we can go back to Erving Goffman's ideas about the theatricality of everyday life (Goffman 1990) and the roles we play, manipulating other people and changing our behaviour in accordance with our interlocutors. We can also see an element of play, both playfulness and theatricality. Thus, in *Photograph 2* Brown is subordinate. But in *Photographs 3a* and *3b: Whiffen and John Brown Dressed as Indians*, he is imitating his former boss dressed as an Indian in the left photo, even outdoing Whiffen, with feathers and a straw skirt. In *Photograph 4: John Brown in Top Hat*, taken by Whiffen, he is dressed as man from the upper classes. And later, in old age in Colombia, he became a kind of tourist attraction, as we saw in McKenna's report, in *Photograph 5: John Brown, with Daughter and Anthropologist Horacio Calle Restrepo*, a *Preto Velho*[44] in *Photograph 6: John Brown, Book Cover*, taken by François Dolmetsch in Puerto Leguízamo in 1964 (Wylie 2013), and in *Photograph 7: John Brown, Very Old.*

---

[44] According to Wikipedia in Portuguese: "Pretos Velhos or Pretos-velhos are a form taken on by Umbanda entities. They are spirits that present themselves under the archetype of old Africans who lived in the slave quarters, mostly as slaves who died in the stocks or of old age, and who love to tell the stories of the time of captivity. They are considered purified deities. Wise, tender and patient, they give love, faith and hope to "their children" [...] The Preto Velho is also well-known for his pipe".["Pretos velhos ou Pretos-velhos são uma linha de trabalho de entidades de umbanda. São espíritos que se apresentam sob o arquétipo de velhos africanos que viveram nas senzalas, majoritariamente como escravos que morreram no tronco ou de velhice, e que adoram contar as histórias do tempo do cativeiro. São considerados divindades purificadas. Sábios, ternos e pacientes, dão o amor, a fé e a esperança aos "seus filhos" [...] O Preto velho é lembrado também pelo instrumento que normalmente utiliza, o cachimbo].

CHAPTER 7

THE SOLDIER –
VERNON (DICK) WALTERS:
MR. UNDERGROUND

1. The Military Attaché

One of the most important and influential interpreters is Vernon Anthony (Dick) Walters (1917-2002), interpreter of many of the most important figures of the second half of the 20th century such as US presidents Truman (1945-1953), Eisenhower (1953-1961), Johnson (1963-1969), Nixon (1969-1974), Ford (1974-1977), and Reagan (1981-1989). He worked with George H. W. Bush at the CIA as Deputy Director when Bush was Director (1976-1977). He also came into contact with almost all the leading actors on the international stage in the second half of the twentieth century, with some as interpreter, and with others on secret missions: George Marshall, Averell Harriman, General Douglas MacArthur, Winston Churchill, Général de Gaulle, General Franco, Konrad Adenauer, Helmut Kohl, Margaret Thatcher, Georges Pompidou, François Mitterand, Juan and Evita Perón, Fidel Castro, Marshal Tito, President Zia ul-haq of Pakistan, Field Marshall Montgomery, King Hassan II of Morocco, and, last but not least, Popes Pius XII (1939-1958), John XXIII (1958-1963), Paul VI (1963-1978), and John Paul II (1978-2005), for whom he was Ronald Reagan's intermediary and, possibly with the help of the Pope, helped to undermine General Jaruzelski's communist régime in Poland (1981-1989) (Jackson 2002). The cover of the 2003 edition of Walter's memories, *The Mighty and the Meek*, shows Walters beaming between President Richard Nixon and Pope Paul VI[45].

---

[45] The most bizarre thing I discovered when writing this book was the existence of a punk band called The Vernon Walters, active in the Netherlands between 1986 and 1990.
https://www.google.com/search?q=you+tube-The-Vernon-Walters&oq=you+tube-The-Vernon-Walters&aqs=chrome..69i57j69i64.2625j0j4&sourceid=chrome&ie=UTF-8.

He was the interpreter at some of the key events of the 20th century. He accompanied Averell Harriman to Europe in 1948 for the implementation of the Marshall Plan. He was Eisenhower's interpreter on the tour of Europe in January 1951 to visit several countries to establish SHAPE (Supreme Headquarters Allied Powers Europe) in Belgium and NATO (Walters 1978); and he was also at other NATO-related meetings. He was President Eisenhower's interpreter when he managed to improve relations with Général de Gaulle on his visit to Paris in 1959; he was the interpreter for Henry Kissinger on secret visits to Paris in 1969, 1970, and 1971, which Walters himself organized, when Kissinger met with the North Vietnamese to discuss ending the war, establish relations with China, and plan the visit of the President Nixon to China in February 1972. Indeed, the Chinese told George H. W. Bush, director of the CIA when Walters was his deputy director, that Walters helped "open the door" (Walters 2000:87).

Walters was the son of an English insurance salesman, Frederick Walters, and his mother, Laura O'Connor Walters, was American. He was born in New York City on 3 January 1917. When he was six years old, his family moved to Europe, where Walters completed all of his formal education at the private Jesuit school in Paris, Saint-Louis de Gonzague, and then at the top Jesuit boarding school, Stonyhurst College, in Lancashire, England. In Paris Walters learned to speak French like a native; he also learned Italian, Spanish, German, and later Russian, Polish and Dutch. In 1933, back in the United States, his father faced financial problems, and young Vernon left school to work with his father, initially as an office boy, and later as an insurance adjuster and investigator, in many cases dealing with clients who spoke the languages in which he was fluent. He never attended university (Murawiec 1984:51-52).

On 2 May 1941 Walters enlisted in the army as a recruit, and the following year he attended Officers' School, where he graduated as a Second Lieutenant in the infantry. In World War II he participated in Operation Torch, the landing in Safi, Morocco, on 8 November 1942, where he worked as an interrogator of captured French soldiers who had been fighting for the Vichy puppet government. Interestingly, many of these soldiers ended up fighting for the Free French. In 1943 Walters had to act as a guide for a group of Portuguese and Brazilian soldiers visiting the United States. He still didn't speak Portuguese, and in order to fulfill his mission he had to learn the basics of the language in a few days. Two years later he served in Italy as a Liaison Officer between the 5th Army and the Brazilian Expeditionary Force (FEB), where he became friends with several Brazilian officers and was promoted to Major in September 1944. After the conflict ended, in 1945 Walters was appointed assistant to the Military Attaché at the US embassy in Brazil, later moving to Paris in 1948, where he was Attaché General. From 1950 to 1955 he was executive officer of SHAPE in Europe and was appointed Aide-de-camp to Averell Harriman (1891-1986), then responsible for implementing the Marshall Plan. On 15 October 1950 Walters attended the famous meeting on Wake Island between President Truman and General MacArthur (1880-1964), which would later result in Truman's firing MacArthur (Montgomery 2004:9-12).

While a staff assistant to the Pentagon and the White House from 1955 to 1960, he worked as an interpreter for Richard Nixon in 1958, on the Vice President's tour of seven countries in South America: Uruguay, Argentina, where he attended the inauguration of the democratically elected,President Arturo Frondisi (1908-1995), Bolivia, Peru, Ecuador, Colombia, and Venezuela. The tour was ironically called the Goodwill Tour, but in several countries there were protests and demonstrations against the USA: in Montevideo on arrival, and in Lima when he deposited a garland on the monument to San Martín (Nixon's Tour:Internet). In Lima he was also prevented from delivering his speech at San Marcos University. There was much criticism in Latin America of the Eisenhower government for supporting dictatorships and the military rather than economic projects. In Caracas protesters approached the car that picked up the politicians on the airport tarmac, and Nixon, Walters, and Mrs Nixon were showered with spit by the protesters upstairs as they entered the terminal. Apparently much of the anger against the United States was due to the fact that the country gave asylum to former dictator Marcos Pérez Jiménez (1914-2001), deposed in January 1958. The following photos are from the Venezuelan leg of the Goodwill Tour.

Photograph 1: Walters Interpreting Nixon

Photograph 2: Nixon Speaks

Let us make an initial analysis of the two photos above, in which we see a clear partnership between the civilian and the military. Nixon had the option of using an official White House interpreter, but he preferred to take Walters with him. It was the first contact Walters had with Nixon, but he had already translated many times for President Eisenhower. According to Harry Obst, official German interpreter at the White House, there was a certain tension between Nixon and the group of interpreters: "suspicion and mistrust were the hallmark of this president and his inner circle" (Obst 2010:70). "Nixon kept himself at a cold distance, and some of my colleagues told me that they now felt seen as if they were walking dictating machines" (Obst 2010:70), and "As we were now largely outside in the cold, trusted friends White House or foreign interpreters used to take our place" (Obst 2010:70), Walters being one of these outsiders who took the place of official interpreters.

Photographs 1 and *2* are by Paul Schutzer (1930-1967) for *LIFE* magazine. *LIFE* was the most important photojournalism magazine at the time, with a print run of around one million copies. *LIFE* prized the quality of its photographs, and Schutzer would also work in Vietnam, where he took the photo for a famous *LIFE* cover, "The Blunt Reality of the War in Vietnam", dated November 26, 1965, showing a Vietcong prisoner with his eyes blindfolded and a tape covering his mouth.

LIFE photographs fall into the category of documentary photography, a *documentum*, a medieval term for an official paper, evidence of what happened, or even a true version with the authority of law, and in the case of photography, showing what happened. Before the arrival of television, newspapers and especially magazines such as *Picture Post*, *Time*, and *LIFE*, as well as newsreels such as *Pathé*, visually depicted events in the news. According to Graham Clarke, documentary photography

> assumes a bond between reader and subject, buoyed up by an assured mandate not just to record, but to expose: the camera with a conscience [...] the photograph, as evidence of events, was basic to the presentation of the story. In terms of the twentieth century, documentary photography has visualized history as a series of events and discrete images which speak of the complexities of human experiences and disaster (Clarke 1997:145).

It is in documentary photography that the cliché "a photograph never lies" is strongest, and the way in which

> the photograph has been understood through its ability to record an objective image of event with an assumed veracity that painting and drawing could

never claim with equal authority [...] The trace of the past, the mark of historical significance, clings to such images, giving them an almost talismanic quality and presence as evidence of what was (Clarke 1997:145-146).

For Elizabeth Edwards "Photography brings the expectancy of the real, the truthful. The immediacy and intimacy offered by the photograph also suggest 'truth' for intimacy and truth are perceived as largely contingent on one another". However, photographs resist this "truth" as in the words of Siegfried Kracauer, "they do not lend themselves to being dealt with in any definite way" (Kracauer 1995a:191-2 in Edwards 2001:10).

Documentary photography purports to provide a "neutral" image, but "It ignores the entire cultural and social background against which the image was taken, just as it renders the photographer a neutral, passive, and invisible recorder of the scene" (Clarke 1997:146). But, of course, this "neutrality" must be questioned. In *The Greatest Magazine Ever Published*, David Plotz summarizes *LIFE*'s view of the world after World War II:

> Fresh off their triumph in the war, their political ideals vindicated, their science and engineering proven the best in the world, Americans were happy to be flattered by *LIFE*'s positive, optimistic vision of post-war society. [...] The story *LIFE* was telling about America wasn't a true story – midcentury America without racial conflict is a fantasy – but it was truish, and it was a story that many Americans wanted to hear (Plotz 2013:Internet).

Clive Scott compares the documentary photographer to the street photographer. The work of the documentarist will pursue a "rhetoric of completeness, so that evidence can become proof". It is always indexical, referring to fixed qualities, here the strength, the toughness of the representatives of the USA. It works with moral abstractions: the freedoms, solidity, democracy of the US way of life. By contrast, the street photographer is more interested in the unpredictability of human agency, pursuing "behavioural characteristics in action" (Scott 2007:72).

In fact, the *LIFE* photos showed the ideals of the American vision. Roland Barthes mentions that the photographs of the Hungarian André Kertész were rejected by *LIFE* when he arrived in the United States in 1937 because, with their unusual angles and unorthodox compositions, "they spoke too much" (Barthes 1981:38; Hill and Cooper 1992:45) and would not fit in with the optimistic vision of *LIFE* magazine, which wanted "documents, techniques, not expressive photographs" (Hill and Cooper 1992:45).

Photographs 1 and *2* reinforce the vision of US strength and superiority, which we will now describe in our analyses. In *Photograph 1: Walters Interpreting Nixon,* we see Walters interpreting Nixon from memory, unlike the vast majority of professional interpreters, who would use a notebook. They are in the open air, under an awning. Nixon appears calm, perhaps trying to understand Walters' Spanish, as he had some knowledge of the language, or perhaps dreaming, looking at a point in the distance, the vanishing point. The parts of the face that stand out are the jaw, mouth, and upper lip. Nixon has a prominent, square chin, associated with traits like "strong, stubborn, honest, down-to-earth and can survive in the harsh environment" (Chin/Jaw Face Reading:Internet), and a mouth with "a stiff upper lip". The expression "to keep a stiff upper lip" is traditionally associated with the English attitude of remaining resolute and determined, and showing no emotion when confronted with adversity. Walters looks down as he translates, not wanting to dominate the scene, although his translations were always well received, and, in some cases, as we shall see, they were considered better than the original speech. The microphone must be very powerful as Walters is quite far away. It is from the WRUL (World Radio University Listeners), "a non-commercial station with sponsorship owned by Worldwide Broadcasting Foundation" that transmitted The Voice of America, the US international news service (The Global Medium on the Massachusetts Coast:Internet).

In *Photograph 2: Nixon Speaks*, we have a clear view of the cloth covering the table, with the words "Radio Libertador", and its frequency. As the photos we found on the Internet do not give the location of the photos, only stating that they are from Nixon's tour in South America, only later could we discover that we are in Venezuela. Again, we note that Walters does not use any kind of notes to interpret. In both photos, in the background we see several men whose age must vary between 30 and 60 years old, some with Anglo-Saxon features, others with more Latin features. None of them appear to be security or bodyguards for the Vice President.

We can see certain anomalies in the photos. Nixon's tour of seven countries in South America was called the Goodwill Tour. In this case, wouldn't it be more appropriate for Walters not to wear his military uniform? Above we saw Nixon's lack of trust in the White House permanent interpreters and his preference for Walters, but might not the prominence of Walters in uniform convey a message that was far from "goodwill"? Isn't this a visual display of US military might? A warning to the countries in the US "backyard" to behave themselves? Or perhaps for bureaucratic reasons, since he was a member of the army on duty. However, the fact that Nixon was accompanied, interpreted, minded by an army officer, well over six foot tall and burly, could be considered a provocative element by many, with the obvious visual association between US presence and military strength. But could we also interpret the union of Walters and Nixon as a way to demonstrate the support that the US military has for the civilian government? Perhaps to show that in the US the military always follows and obeys the civilian government, staying, like Walters in the photos, one step behind. The initial vision given by the two photos in *LIFE*, which was always a highly patriotic magazine, especially during the Second World War, is that of two strong, unflappable men, representing the power of the United States, who would support their friends but who would defend themselves with strength and military power against their enemies.

Finally, a *punctum* (Barthes 2015:27-31 passim): why is Walters uniform so wrinkled? Was there no one to iron it before he got on the platform?

2. By a Whisker…

They arrived in Caracas on 13 May 1958. It was decided that from the airport they should travel to the US Embassy in an armoured limousine, and not in a convertible, as had happened in other countries. On the runway at the airport they were surrounded by a crowd protesting and spitting on them from a balcony of an airport building. The images can be found on the Internet. On the road from the airport traffic stopped because of the demonstrators who had blocked the way and who started hitting the armoured car. The twelve American security guards tried to hold back the demonstrators, who managed to break the windows of the armoured limousine and rock the car, but Nixon asked the security guards not to shoot.

We can read the report from *LIFE* reporters, who were following the tour:

> A rampant mob, befouling the Vice President of the United States and his wife with insult and filth and trying to kill them, last week brought an eight-nation "goodwill" tour to a lurid climax […]. "They were out to get Nixon from the minute he stepped off his plane", cabled *LIFE* Correspondent Donald Wilson, "firstly hecklers, then screams and spit as they went to their car" (*LIFE* 1958:32).

When the car stopped on Avenida Sucre:

> They began pounding on the door with sticks. The escort, a handful of Venezuelan police, stood by listlessly. Nixon sat tight-lipped as broken glass flew like snow. Venezuelan Foreign Minister Oscar García Velutini threw

his hands up as glass hit him in the face. A grapefruit boulder crashed into a window, stuck, then dropped to the floor. The mob tried to overturn the car (*LIFE* 1958:32).

Some North Americans accompanying the visit looked on closely without being able to do anything. "I felt at once quick anger and a swelling urge to cry", reported *Time* and *LIFE* correspondent Bruce Henderson. "I was an American and here before my eyes the Vice President of the U. S. was on the verge of very possibly being beaten to death. How in God's name could something like this be happening in our own hemisphere backyard?" (*LIFE* 1958:32)

Apparently, the police did nothing to help the Americans. Several members of the entourage were cut by the glass, and nonchalantly Nixon told Walters, "Well, you're bleeding from your lip, spit out that piece of glass, because I have a few more things to say to these Venezuelans today" (Walters 2000:62; and Rutler 2005:Internet). Finally, after about twelve minutes, the police managed to break through, and, instead of going to the Simon Bolivar Liberator monument to leave a wreath, they went directly to the US Embassy, where the above photos were taken. For Pathé News it was the most violent attack ever perpetrated on a US official on foreign soil" (British Pathé News:Internet), and Nixon himself said: "I felt as close as anyone could get to death and still remain alive" (The History Guy:Internet), a feeling probably shared by Walters, who throughout his military career had never entered a battlefield.

Upon learning of the incident, President Eisenhower sent aircraft carriers and marines to Puerto Rico, in readiness to invade Venezuela in the event of further violence to the Vice President. However, the Venezuelan army was mobilized to ensure that there were no more demonstrations, and the next day, after meeting with the military junta that governed Venezuela, they brought the return flight forward by twelve hours and arrived at the airport without any problems. After a stopover in Puerto Rico, they reached Washington on 16 May. Nixon was given a hero's welcome in Washington; federal officials were given a day off; and 100,000 people cheered Nixon on the way from the airport to the White House. However, Eisenhower's provocative reaction was highly unpopular, and the Republican administration's foreign policy was one of the reasons why Nixon lost the presidential election to John Kennedy and the Democrats in November 1960. Walters never worked for Kennedy.

With this more detailed knowledge, let's return to *Photographs 1* and *2*. The interpreter Walters was almost ignored by *LIFE,* similar to the way in

Chapter 1 in which Arnold Vissière was ignored by the newspapers in the press reports on the Chinese delegation's visit to Paris. Here Walters was only mentioned in the caption for *Photograph 1: Walters Interpreting Nixon*, which was published in *LIFE*. From *LIFE* we know that one of the things Nixon said in the speech was: "It is not easy to endure the kind of activity we had to go through" (*LIFE* 1958:32). *LIFE* also comments on a photo with Pat Nixon, wife of the Vice President, talking to other women at the embassy, just an hour after the ordeal. In *Photo 1* we can imagine that Nixon is thinking about the recent incident and how lucky he is to be alive. And Walters had no time to have his wrinkled uniform ironed after arriving at the Embassy, or perhaps he was too traumatized to bother!

3. After the Goodwill Tour

In May 1960 Walters was appointed Military Attaché in Rome, at a time when the CIA was trying to prevent socialists and communists from taking power in Italy. In October 1962 he was appointed Military Attaché in Brazil, at the request of the US ambassador to Brazil, Lincoln Gordon (1913-2009). He was promoted to brigadier general in February 1965 and remained in Brazil until June 1967, when he spent a short time in Vietnam with US troops supporting the Saigon government in the war against North Vietnam and Vietcong guerrilla forces in the south of the country.

Although Walters wanted to stay in Vietnam, in August 1967 he was transferred to Paris, again as Military Attaché, and stayed there until 1972, accompanying President Nixon on tours of various countries in 1969, the Azores Islands in 1971, and Mrs. Nixon to the regions of Peru devastated by an earthquake in June 1970. In February 1968 he was promoted to Major General and attached to the Undersecretary for Southwest Asian Affairs, Averell Harriman, with whom he had already worked in the implementation of the Marshall Plan. At the end of that deployment, he was promoted to Lieutenant General in April 1972. Between May 1972 and 1976 he was Deputy Director of the Central Intelligence Agency (CIA) under directors James Schlesinger, William Colby, and George H. W. Bush, and was Acting Director from 2 July to 4 September 1973, right around the time of Watergate. It is possible that his refusal to cover up the Watergate investigation resulted in the investigation that followed.

In 1976 he officially retired and was not called to work in the Jimmy Carter 1977 to 1980 administration. However, from 1981 to 1985, under President Ronald Reagan, Walters served as a roving ambassador. He was then US Ambassador to the United Nations from 1985 to 1988 and Ambassador to

Germany from 1989 to 1991, during the time of German reunification. Walters is a member of the Military Intelligence Hall of Fame.

4. Mister Underground

Above we have the official version of Walters' career, but there is also the unofficial version that tells a somewhat different story. On the surface Walters was a paper-clip general, a mere bureaucrat:

> Like any paper-clip general, he has never demonstrated any burdensome knowledge of military science. But he knows all about back-stabbing, lying, cheating – and coup – plotting, assassination, destabilization, and disinformation operations, for all of which, at one time or another, he has been a case-officer (Murawiec 1984:51).

He was also called Mr. Underground (Oliveira 2005) due to his mania for knowing all the underground railway networks in the world, and his characteristic of trying to find out what was happening in all the places he visited in an "underground" way.

Mixing duty and pleasure, Walters used to visit in advance the country in which he would take part in important meetings. His aim, when riding the bus and metro, was to remember local slang and accents, in addition to gathering people's impressions regarding subjects of interest to the United States of America. His liking for the underground was strategic (Oliveira 2005:45).

One example is from 1968, during the student riots in Paris. In an interview in *El País* he describes his way of finding out what was going on: "Every night, right after the riots started, I would go to the eastern neighborhoods of Paris and drink wine with the workers, with some policemen and some soldier"[46] (Espada 2000: Internet).

Father George W. Rutler gives a more sympathetic portrayal of Walters:

> I recently asked Kissinger to summarize Walters' diplomacy. "Flamboyant discretion", he instantly replied; "Dick" Walters loved the drama of it all. As a teetotaling, non-smoking, chaste bachelor, he was a kind of ascetic James Bond, with the added advantage that he was real (Rutler 2005:Internet).

---

[46] "Todas las noches, a poco de empezar los motines, iba a los barrios del este de París y tomaba unos vinos con los obreros, y con los policías, y con algún militar".

Perhaps more of a Forrest Gump than James Bond, turning up everywhere where something was happening! Murawiec is less fawning, pointing out the sinister, manipulative side of Walters and his contacts with conservative Catholic groups. He was the

> Jesuit lapdog controlling George H. W. Bush while Bush was director of the CIA (Catholics in Action) – being Bush's deputy director. Walters was very valuable to the Jesuits and seemed to be in many locations where CIA backed coup d'etats (sic) were taking place" (Murawiec 1984:51).

Murawiec also comments on Walters' asceticism:

> Walters lived with his mother throughout his adult life until her death; he now lives with his sister. Gossip about his sexual proclivities was laid to rest by an authoritative Washington source who inhaled deeply, and pronounced the word "eunuch". He is generally described as a "fat pompous ass" (Murawiec 1984:51).

Walters made a number of friendships with figures who would be important in their own countries. In 1942 the thirteen-year-old boy, the future King Hassan II of Morocco, whose rule was aided by the United States, climbed into Walters' tank in Morocco (Walters 2000).

Walters was Averell Harriman's interpreter for negotiations in 1953 with the Iranian leader, Mohammad Mossadegh (1882-1967), to try to prevent him from nationalizing the Anglo-Persian Petroleum Company, now British Petroleum (BP). Subsequently the Mossadegh government (1951-1953) was overthrown and replaced by Shah Mohammad Reza Pahlavi (1919-1980), whom Walters had met during negotiations (Murawiec 1984). He participated in several US actions in other countries. In 1965 Walters negotiated Brazil's participation in the US invasion of the Dominican Republic, an agreement signed by Harriman.

While in Paris he was loaned by Harriman to Kissinger and entrusted with the secret transfer of Kissinger to meetings with the North Vietnamese from 1969, and with the Chinese, a total of fifteen clandestine meetings between 1970 and 1971, without the knowledge of the CIA (Walters 1978:576-583).

He was part of a mission in 1973 to try to convince the PLO (Palestine Liberation Organization) to stop killing Americans. He also participated in a mission to Paraguay in 1976, before leaving the CIA, to free a US spy who had insulted President Stroessner. In 1978 he participated in a mission to Pakistan to obtain, through a large sum of money, the Russian M-24 plane brought by a fugitive pilot from Afghanistan, when the Soviet Union was fighting the Mujaheddin in Afghanistan. Another mission was in 1984 when he went to Ethiopia to obtain the release of an American prisoner. In March 1984 he persuaded the Islamic president of Sudan, Jaafar Nimeiry, not to impose sharia law in the Christian south Sudan. In 1985 he travelled to Nigeria and convinced the army chief, General Buhari, who controlled the country, to allow the US to use the Port Harcourt to transport food to Mali and Niger and to release Marie Lee McBroom, a black American trader arrested for violating a law related to the oil trade. In 1986 he persuaded President Ceaucescu of Romania not to tear down the Jewish museum and allow the importation of bibles. And between 1980 and 1986 he made six visits to President Assad of Syria to finally secure the release of Hezbollah hostages (Walters 2000).

It is also likely that he was involved in several of the CIA-sponsored bloody coups: Iran in 1953, Guatemala in 1954, Brazil in 1964, and Chile in 1973, when he was Deputy Director, as well as the establishment and training of General Pinochet's secret police in Chile. Walters always defended the presence of the military in Latin American governments: "The Latin American military are a stabilizing force and a block to the ambitions of the Communists" (Walters 1980:298), and "don't denigrate right-wing dictatorships: They almost always lead to democracy – Portugal, Spain, Greece, Argentina, Brazil" (Weaver 1986). He may also have been involved in the Falklands War, deceiving Argentina after the invasion of the islands by assuring that the US would not oppose the invasion. The opposite happened, with the US giving considerable support to the UK (Murawiec 1984:52).

And, according to Murawiec, between 1976 and 1981 Walters was officially "retired" but was paid high salaries for "consulting" positions, including $300,000 from Environmental Systems, Inc., which exported weapons and tanks, together with other military equipment (Murawiec 1984:52).

In *Le Cercle and the Struggle for the European Continent: CIA, MI6 and Opus Dei Covert Politics* (2018), Joël van der Reijden, a researcher at the Institute for the Study of Globalization and Covert Politics, details Walters' involvement with far-right informal Catholic groups such as Le Cercle, Gladio, 61, Condor, Safari Club, which often had links to the Vatican, European aristocracy, and bankers.

Walters participated in the nefarious National Security Council (NSC), helped found the Safari Club, and became a member of the Sovereign Order of Malta in 1977. He established 61, a "private" international agency created in 1977 with Brian Crozier "to delude the official intelligence services". He was also one of the founding members of the Safari Club along with Count Alexandre de Marenches, head of SDECE (Service de Documentation Extérieure et de Contre-Espionnage), and they recruited Anwar Sadat, President of Egypt, Saddam Hussein, and the Shah of Iran for this anticommunist alliance, and George H. W. Bush helped bring in the Saudis.

He was one of the founders of Operation Condor, which operated in Latin America as a kind of anti-socialist death squad, with the backing of the CIA. Reijden (2018) details Walters' movements: in the summer of 1974 he was in Chile to talk with Pinochet, who had recently taken office; he was a friend and had a lot of contact with the head of Pinochet's security services, General Manuel Contreras (Executive Intelligence Review 2005:Internet), and the FBI suspected that Walters had some connection with the car bomb that killed Chilean refugee Orlando Letelier on 21 September 1976 (BBC 2005:Internet).

From 1977 to 1980 Democratic President Jimmy Carter did not call Walters to any kind of assignment, but with Walters' appointment as roving ambassador under Ronald Reagan, he was a "key emissary of the covert channels" in Zaire, Kenya, Morocco, Ceylon, India, Nepal, Angola, El Salvador, Argentina, Zambia (Boston Globe 1982). Some of the missions have been reported above.

In 1981 he twice visited Guatemala, now under the control of right-wing Romeo Lucas García, and restored financial aid to Guatemala that had been cut off by Carter. He organized support for the Nicaraguan Contras, supported by the Reagan administration. And Reijden says that in May 1984 Walters convinced Roberto D'Aubuisson of the Contras not to assassinate the US ambassador in El Salvador.

The next section analyzes Walters' role in the 1964 military coup in Brazil.

5. Walters and the 1964 Military Coup in Brazil

President João (Jango) Goulart's government was moving further towards the left, and many worried that Brazil would follow Cuba into communism and dictatorship. The United States were very worried, and Walters as Military Attaché was in a key position. In relation to his participation in the 1964 political events in Brazil, Walters, now Colonel, stated:

> I was the military attaché and I didn't participate in any conspiracy. I was
> an informed witness but not a participant. Personally, I was very worried
> about President João Goulart's rally on March 13, 1964, in Rio de Janeiro,
> with the red flags... I was a foreigner and had lived in Brazil for some years...
> I had the right to observe, but not to participate (Contreiras 2002:Internet).

Others take a different view. In *O Governo João Goulart: As Lutas Sociais
no Brasil* (1961-1964) [The João Goulart Government: The Social Struggles
in Brazil (1961-1964)] Luiz Alberto Moniz Bandeira, a Brazilian historian,
supports the fact that Walters played a more active role in the episode, not
only coordinating CIA activities in the country, but also participating
directly in the planning of the coup (Bandeira 1978:48).

In the documentary *O Dia que Durou 21 Anos* [The Day that Lasted 21
Years], by Camilo Tavares, Walters' central role is clear. We hear the voice
of Walters advising the US ambassador, Lincoln Gordon, a few days after
the coup:

> I think the inauguration of Castelo Branco in civilian clothes was significant.
> With his desire to emphasise the civilian character of the government. This
> is fully supported by the army, which has great confidence in him, as I think
> very broad sectors of Brazilian opinion do (Tavares 1971:58-59 min).

In fact, Walters was a great friend and admirer of the Brazilian general:

> […] I never saw Castello (sic) Branco have a less dignified attitude or utter
> a foul word. This man's moral integrity was beyond provocation. In the
> midst of danger, he clearly demonstrated that, like all of us, he felt fear, but
> the iron self-discipline he always maintained his undisturbed calm (Walters
> 1986:121).

In the documentary Peter Kornbluh, Coordinator of the National Security
Archives, clarifies Walters' role:

> General Vernon Walters obviously did reports on all of the Brazilian
> military figures he was dealing with. He certainly told his superiors not only
> in the US military but also in the State Department and the White House the
> most efficient, who he thought was the most pro-US, the most professional,
> the most likely to be an efficient and strong coup leader (Tavares
> 1971:36m).

James Green, historian at Brown University, continues:

> As his mission was basically to organize the conspiracy by bringing together
> the different dissident groups within the army which were interested in
> doing something to stop or to overthrow the Goulart government, so he went

patiently from one person to the next, meeting with different colonels and generals, assuring them that the US would support the military if it were to take power (Tavares 1971:36m).

The documentary switches back to Kornbluh: "and he picks out Castelo Branco", showing a copy of a telex from Walters:

> By all odds the most significant development is the crystallization of the military resistance group under the leadership of General Humberto Castelo Branco, honest, and an Army Chief of Staff, he is highly competent, discreet and respected officer who has strong loyalty to legal principles (Tavares 1971:36m).

The support of the USA and the central role of Walters is clear for Bandeira, who mentions the deposits of war material hidden in churches and farms (Bandeira 1977:126); the contact that Walters had with the armed forces, especially through his contacts at the Escola Superior de Guerra (Bandeira 1977:130); the large number, approximately 5,000, of green berets, the US special forces, who were "visiting" Brazil on "non-war" missions; and Operation Brother Sam, with a large task force, consisting of a carrier planes, destroyers, ships loaded with marines, supplies, fuel and weapons, ready to enter Brazil in case of the need for intervention (Bandeira 1977;174). Bandeira recounts that it was Walters who received the information that the sedition of the Minas Gerais forces would take place on 31 March through the collaboration of the CIA with various currents of opposition to Goulart (Bandeira 1977:176). And it was possible that Walters reported this fact to Castelo Branco when he had dinner at his house on the night of 30 March 1964 (Bandeira 1977:137-138). The reports of the SFICI (Federal Information and Counter Information Service) of the Goulart government mentioned his meetings with a large number of military personnel. Citing the interview with General Guedes, "Walters, who even offered them war material, *tried to open their eyes*, in other words, instigated them to rise up against the government"[47] (Bandeira 1977:151). Walters, with his excellent Portuguese and his friendships with a large number of military personnel, reined in the coup and tried not to give the impression that the CIA was behind it. Quoting Bandeira's words again: "In Goulart's opinion, 'the technique used was perfect'. The CIA handled the situation discreetly, as Walters believed 'that Brazilians would be offended if the

---

[47] "Walters, que lhes oferecera até material bélico, *tentava abrir-lhes os olhos*, em outras palavras, instigava-os para que se insurgissem contra o governo".

United States showed leadership in the coup'"[48] (Bandeira 1977:166; and Parker 1977:94). According to Phyllis Parker, the United States did not get involved in carrying out the coup as there was no need to (Parker 1977:129). It was not necessary to use the weapons, marines, and oil stock, the aircraft carrier Forrestal and support destroyers, which were sent towards Brazilian waters when the coup was triggered on 31 March.

Using US documents and interviews with the main actors, Phyllis Parker, in *1964: O Papel dos Estados Unidos no Golpe de Estado de 31 de Março* [*Brazil and the Quiet Intervention, 1964*] (1977), agrees with Bandeira regarding the role of Walters, whom she interviewed on 20 January 1976. Walters told her that Castelo Branco had joined the conspirators in January or early February 1964 and became their leader (Parker 1977:83); Castelo was also concerned about the rally that Goulart called on 13 March, when which he told Walters that the only symbols he had seen were hammers and sickles (Parker 2014:101).

In the interview, Walters emphasizes his own role as intermediary, telling Parker that it was he who informed the US ambassador, Lincoln Gordon, that Castelo Branco had assumed leadership of the movement against Goulart (Parker 2014:129-130).

As noted above, Walters had regular meetings with Castelo before the coup but avoided contact with him on the night of the beginning of the revolution, 31 March, trying to be more discreet, when General Olímpio Mourão Filho was already moving the Fourth Army of Minas Gerais, unknown to Castelo Branco. However, Walters went to the apartment of General Floriano de Lima Brayner, who informed him that General Amauri Kruel, commander of the Second Army of São Paulo, had given clear support to the rebels, at a time when the situation was not at all clear (Parker 2014:151). Walters then sought Castelo again at two o'clock in the morning on 1 April, and again on the morning of 2 April, when it was clear that there would be no resistance from Goulart (Parker 2014:155).

In *Photograph 3: Friends and Brothers*, we see the very tall Walters (almost two metres) and the short Castelo Branco (1.64m) looking at each other with their profiles facing the camera, in a corridor or lobby of a building. An obvious interpretation of the photograph would be to see the big, burly Walters as representing US power and hegemony, dominating small, weak

---

[48] "Na opinião de Goulart, 'a técnica usada foi perfeita'. A CIA manejou discretamente a situação, pois Walters acreditava 'que os brasileiros se melindrariam se os Estados Unidos demonstrassem liderança no golpe'".

Brazil. However, the two are smiling, and, as we know, they have been friends for twenty years, since World War II. So, rather than seeing the photo as representing the great power of the USA dominating the relatively weak Brazil, beset by economic problems, the two are in complete harmony. Little Brazil is very friendly with the big USA and does not feel intimidated, and they will face the future together and in friendship.

Photograph 3: Friends and Brothers

It is interesting that neither is in military uniform, pointing out that the problems will be resolved by civilians in peace and harmony, after the turbulent years of Goulart, with no need for any further military intervention.

We do not have the date of the photo, but it is from the Centre for Research and Documentation of Contemporary History of Brazil (CPDOC) of the School of Social Sciences of the Getulio Vargas Foundation and the Jornal do Brasil and Folhapress, and it is almost certain that it was taken after military coup as before Walters needed to hide his proximity to Castelo.

5. Walters Interpreter

Walters comments little on his interpreting style and technique. Possessing a prodigious memory, a native French speaker and fluent in Spanish, Portuguese, Italian, German and Russian, he was an intuitive interpreter. He never took notes, never took an interpreting course and was capable of consecutively interpreting long speeches.

However, in *Secret Missions* he makes a number of comments. Interpreting for Général de Gaulle in Italy in 1944, discussing the withdrawal of French troops who were to participate in the landings in southern France under the command of the American General Clark, then Walters' superior, and thinking that de Gaulle would not understand English, he adds some comments for General Clark, for example, "Général de Gaulle says he doesn't agree, but I think if you insist a little he'll give in eventually", or "He said he agrees, but I have the impression that he is not very convinced". At the end of the session de Gaulle greets him in good English: "Walters, you did a great job". Walters is contrite: "it served as a lesson to me never to meddle in the subject matter of translation. This happened to me once, later, with the Brazilians, also during the war" (Walters 1986:96-97).

Another occasion when Walters felt compelled to do a "diplomatic" translation was on his visit to Iran along with Averell Harriman to act as interpreter into French and try to convince the Prime Minister of Iran to compensate the Anglo-American company Anglo-Iranian Oil Company for the nationalization it intended to carry out. The US Ambassador's wife, Mrs Lucretia Grady "very given to queer ways of expressing herself, at one point said to the Prime Minister: 'Dr. Mossadegh, you have a very expressive face. Whenever he's not thinking about anything, I can tell by the blank look on his face'". Walters, who was trying hard to negotiate with Mossadegh, changed his speech to something more mild: "Dr. Mossadegh, you have a very expressive face. Whenever you are deep in thought, I can see it by the expression of concentration on your face" (Walters 1986:233). However, this was one of the few missions in which Walters was unsuccessful.

On a visit to Bogotá in April 1948, there was an attempted military coup, and at the conference General Marshall was speaking at the simultaneous interpreting equipment had been damaged and, with the riots, the other interpreters were unable to arrive, so Walters was forced to interpret for Spanish, English and French.

Walters tells of several incidents during his visit with Vice President Nixon to South America in 1958, when his car was attacked in Venezuela, as described above. He did not travel to South America with Nixon, and shortly after arriving in Montevideo he was in the Uruguayan Congress translating Nixon's speech, with no paper to take notes and no idea what Nixon was going to say. This was his first contact with Nixon. On the visit to the university, Walters had to interpret the Q&A session with the students, and Walters greatly impressed Nixon (Walters 1986).

In La Paz, Bolivia, Walters had to translate the same speech eleven times. And, on the occasion of receiving the keys to the city of La Paz, an indigenous man began a speech in Aymara. An Aymara speaker whispered a Spanish translation into Walters' ear, and he translated it into English. Nixon and the American newspapers had the impression that Walters also knew Aymara! (Walters 1986:289)

When he was translating Eisenhower's speech at a dinner at the Élysée Palace, on the American president's visit to France in 1959, Walters had to find a solution for: "There are those that say Général de Gaulle is a very stubborn man". In French *entêté* or *têtu* can sound negative, so he used *tenace*: "There are those that say Général de Gaulle is a very tenacious man". De Gaulle had a hearty laugh along with everyone who understood. And this was reported in the French newspapers: "The colonel diplomat covered an ambiguous word with a delicate veil. Once again I had proof of the risk one runs in trying to interfere with the thinking of an authority whom one is serving as an interpreter" (Walters 1986:291-292).

Harry Obst, in *White House Interpreter: The Art of Interpretation* (Obst 2010), shows no sympathy for Walters. Obst was a professional White House German interpreter and trainer of diplomatic interpreters, and he shows that Walters was not well liked, though perhaps envied, among the diplomatic interpreter corps.

Obst found it odd that Walters often interpreted for Eisenhower and Nixon in uniform, drawing salutes from lower-ranking soldiers, a huge distraction for any interpreter (Obst 2010:71). Walters had no training of any kind as an interpreter and did not know how to take notes; in fact, he even found note-taking counterproductive, interpreting five minutes or more from memory. And although he spoke German, Spanish and Russian well, he "never reached the depth of comprehension in those languages that are required of a professional diplomatic interpreter" (Obst 2010:71).

Obst comments on a reception for foreign dignitaries after Eisenhower's funeral in 1969, when Walters was trying to put himself in the right position to interpret for Austrian and German leaders. However, after a skiing accident, he was on crutches, and Obst managed to get in front of him (Obst 2010:72).

Obst comments on Walters' portrayal of President Pompidou when he visited the White House in February 1970. Obst is in the audience at the reception ceremony, and he notes down President Nixon's entire speech, perhaps a little enviously. Walters' French impresses him greatly. As we saw, Walters was a native French speaker:

> and he spoke with clarity and conviction, except an entire sentence and parts of other sentences were missing, several things were changed round quite a bit, and a few words were added that the President never said. However, the main thrust and tenor were there. In fact, Walters' welcoming speech was a little better than Nixon's. Everybody was happy. He probably got a lot of praise for his interpreting (Obst 2010:75).

However, this type of interpretation would not be acceptable for an official White House interpreter: Obst sharply criticizes him: "But if he had done this translation for an interpreter examining board at the State Department, his chances of passing the exam would have been slim" (Obst 2010:75).

Walters was another type of interpreter, the showman, as were several of the League of Nations' famous consecutive interpreters. In his study of United Nations interpreters, *Interpreters at the United Nations: a History*, Jesús Baigorri-Jalón describes the early years of the organization, when consecutive interpreting was still used, and their reluctance to use simultaneous interpreting:

> Consecutive interpreters were accustomed to being among the speakers and to being in a prominent position, because they 'performed' before the audience, from the rostrum. Many of them may have felt that if they were put in a booth with headphones then they were being degraded professionally (Baigorri-Jalón 2004:55).

They were called "mental magicians", "linguistic virtuosos"; they had a reputation for being temperamental and arrogant (Iglauer, Harper's Magazine, April, 1947, in Baigorri-Jalón 2004), and did not want to become "monkeys", "parrots" or "telephone operators", trapped in a booth with headphones (Baigorri-Jalón 2004:55-56).

The following anecdote recalls the style of some of Walters' translations:

> One day a delegate complained about André Kaminker's consecutive interpretation of his speech, saying that he had not repeated what exactly he had said, and Kaminker replied: "I didn't say what you said, but what you should have said" (Baigorri-Jalón 2004:56).

To end this chapter on a high note, perhaps the greatest compliment came from Général de Gaulle, who told Nixon on his visit to the United States in 1960: "Nixon, you gave a magnificent speech, but your interpreter was eloquent" (Appelbaum 2007:Internet).

When showing the photo of Castelo Branco and Walters to groups of students in Brazil, most of whom are postgraduate students, few recognize Castelo, and nobody recognizes Walters. I hope this chapter helps to bring out the great importance of Walters inside and outside Brazil as an interpreter, spy, and the most important North American conspirator of the 1964 coup.

CHAPTER 8

THE "GENERAL" OF THE INTERPRETERS:
VIKTOR SUKHODREV

This chapter examines photographs of the Russian interpreter Viktor Sukhodrev, known as the "general" of international interpreters, and a key figure in the Soviet Union's relations with English-speaking countries, especially the United States.

Viktor Mikhaylovich Sukhodrev (Виктор Михайлович Суходрев) (12 December 1932 - 16 May 2014) was a Russian-English interpreter for high-ranking Soviet politicians, including Nikita Khrushchev (1894-1971) (First Secretary of the Communist Party 1953-1964); Leonid Brezhnev (1906-1982) (General Secretary of the Communist Party 1964-1982); Andrei Gromyko (1909-1989) (Minister of Foreign Affairs 1957-1985); Alexei Kosygin (1904-1980) (Premier 1964-1980); and Mikhail Gorbachev (1931-2022) (General Secretary of the Communist Party 1985-1991). He himself comments in a 2009 Russian TV interview (Russia Today (RT) Spotlight: 2009) that he has also worked with eight US presidents, as well as numerous world leaders, including Harold Wilson and Margaret Thatcher of the UK, Pierre Trudeau and Brian Mulroney of Canada, Lal Bahadur Shastri, Indira Gandhi and Rajiv Gandhi of India, and Pakistan dictator Ayub Khan (Schudel 2014).

1. The English Boy

Sukhodrev's father was a Soviet intelligence officer, i.e., spy (V gostiakh u Dmitria Gordona 2010), in the United States during World War II, but Sukhodrev spent six years in London, arriving there at the age of six, in August 1939, a few days before the start of World War II on 1 September 1939, together with his mother, who worked in the Soviet trade mission. He attended the Soviet Embassy School in London but, living in Carlingford Road, Hampstead, away from the Russian community, he often played with English children in the street and soon acquired fluency in the English language, saying that he grew up as an English child (Sukhodrev 1999).

Thus the young Viktor was part of the group known as Third Culture Kids (or TCK), a term coined by the American sociologist Ruth Hill Useem in the 1950s for children who spend their formative years in places that are not their parents' countries (Baigorri-Jalón and Fernández-Sánchez 2010; Mayberry 2016).

He had a lot of contact with Walter and Catherine Holloway, known as Uncle Jack and Auntie Mary, a couple who lived in the apartment on the floor above his and who often looked after young Viktor, or Vitya. Although Uncle Jack was an engineer by profession, he worked as a postman so he could be outdoors after suffering from gas attacks in World War I. The Holloway couple had no children, and Vitya was like a surrogate son, often accompanying the postman Uncle Jack on his rounds, talking to him and the people who received letters and packages, while learning about local accents and customs. "It was from the postman and his wife, he said, that he learned the British manners that gave him his special grace as an interpreter" (Mydans 2005), and by the age of eight he had already decided to become an interpreter! He says his first interpreting experience was when the Russian school was evacuated to St. Anne's, in northwest England, and he interpreted between the owner of the building company which was doing some repairs to the building, and the school manager, who didn't speak English. And if I was going to do that [for a living], I felt, I was going to be damn good. Maybe the best" (Kokoszycka 2014). He returned to Moscow at the age of twelve, speaking English like a native speaker and later graduated from the Military Institute of Foreign Languages.

2. "We'll show you Kuzma's mother!"

His first experience as an "official" interpreter was when, still a student in 1955, he was part of a delegation of real estate agents from the Ministry of Construction in England (Shevchenko 2020), and he was later promoted by Oleg Troyanovsky, who had worked with Stalin, Molotov, and Gromyko and who was looking for a successor. In 1956 Sukhodrev began his career in the translation office of the Ministry of Foreign Affairs and soon, at the age of 23, became an interpreter for Khrushchev. From 1985 to 1989 he was Deputy Director of the Department for the United States and Canada at the Ministry of Foreign Affairs, and from 1989 to 1994 he was Special Assistant to the UN Secretary General. In 1999 he wrote the memoir *Yazyk Moi – Drug Moi* ["My Language is my Friend", not translated into other languages to date].

During a career spanning nearly thirty years, Sukhodrev was present at numerous high-level summits and negotiations. Richard Nixon said that Sukhodrev was "an excellent linguist who spoke English as well as Russian", while Henry Kissinger called him "unflappable" and a "splendid interpreter" (Schudel 2014). According to the *International Herald Tribune*, "Sukhodrev was present, but not there, emptying himself of the ego, getting under the skin of the man who was speaking, feeling his feelings, saying his words" (Kumiko 2009).

In May 1972 Sukhodrev was the only interpreter during the Moscow summit between Richard Nixon and Leonid Brezhnev: "Despite pressure [...], the President [Nixon] refused to use an American interpreter, relying instead on the Soviet veteran Viktor Sukhodrev, who had translated for Kennedy and Khrushchev in Vienna [in 1961]" (Reynolds 2007). This decision seems somewhat bizarre: with relations still quite tense, Nixon preferred the interpretation of the official Soviet interpreter to the State Department's Russian interpreters, having more confidence in Sukhodrev than in them. Harry Obst, White House German interpreter, comments on the way in which official interpreters were bypassed. Although Sukhodrev was a "brilliant interpreter", his father "had been a Soviet spy in the United States for ten years. How could he be more trusted than an American interpreter [...]?" At the time diplomatic interpreters had to write up the conversations which had taken place from notes and memory. On many occasions Viktor was the only interpreter present, and the Americans had to depend on his accounts. This is no problem to Nixon and Kissinger, but Obst shows great concern that there was no second interpreter to monitor him, and the fact that the Soviet interpreter was also writing the memoranda of conversations for the Americans!

Nixon is frequently dependent on Sukhodrev. He asks Vitya to suggest something short he can say in Russian at a formal dinner, and "Glory to the heroes of Leningrad!" is a great success (Sukhodrev 1999:96); Nixon also asks him to read out the Russian version of a speech he gives on Russian television (Sukhodrev 1999:96); and at a dinner between Nixon and Brezhnev at Nixon's ranch in San Clemente, California in June 1973 Sukhodrev is the only other person present ((Sukhodrev 1999:106). In his memoirs (1990) Nixon later recalled: "I knew that Sukhodrev was a superb linguist who spoke English as well as he did Russian, and I felt that Brezhnev would speak more freely if only one other person was present" (in Kokoszycka 2014). Furthermore, as we saw in the last chapter, it appears that Nixon had a certain fear of the State Department and its interpreters.

In fact, Soviet and American officials considered him the world's best interpreter of Russian and English, and he was often the only interpreter at bilateral meetings. He had a very good understanding of English idioms, with a firm understanding of the various shades of meaning in different parts of the English-speaking world. His memory was prodigious: he only needed a few notes to get a perfect translation of a 20-minute speech. In 2012 Sukhodrev received the Russian National Translator of the Year award.

One of Sukhodrev's successors, Igor Korchilov, writes about the theatricality of interpreting. In Chapters 2 and 7 we use Erving Goffman's concept of the theatricality of everyday life (Goffman 1990). Igor Korchilov (1997 and in Rogatchevski 2009) uses the term *perevoploshchenie*, or identification with the often fictional personality of another person, an important concept in Konstantin Stanislavsky's actor training.

For Martin Kalb, an American correspondent in Moscow, Sukhodrev conveyed Khrushchev's ebullition and excitement and Brezhnev's rather grim and muted conservatism, but would also try hard to smooth out the rough edges and make Khrushchev seem a little more sophisticated and less arrogant than he really was. He also even made Leonid Brezhnev, who wasn't much of an orator, appear more interesting. And Mikhail Gorbachev, who sounded to many Russians like a verbose bureaucrat, became very popular in the West, to a great extent because of Sukhodrev's skills (All Things Considered 2014; Sukhodrev 1999). And Sukhodrev himself, with his American accent (when he was in the United States), speaking "flawless English" (Apple 1973) gave an excellent impression of the Soviet Union itself.

He had the ability to describe sensitive politics and diplomatic manoeuvres with great clarity and could translate slang from one language to another with vivid expressiveness. When Khrushchev was denied permission to visit Disneyland during his visit to the United States in 1959, Sukhodrev interpreted the illiterate Soviet leader's dissatisfaction wittily: "Is there an epidemic of cholera there or sump'n?" (Schudel 2014)

At the famous "Kitchen Debate", the "Kitchen Discussion" at the North American National Exhibition in Moscow on 24 July 1959, Khrushchev was shown the wide variety of North American appliances and a model kitchen. His response was: "Let's show you Kuzma's mother!", which caused some confusion, initially being translated literally, but the North American interpreter ended up making the following translation: "We'll show you what's what".

Sukhodrev wasn't working at this event, but he researched possible translations of the expression: "We're going to finish you off"; "We're going to hit you on the head!" or "We're going to take you to hell!" However, when Khrushchev was driving around Los Angeles in September 1959, he mentioned the expression "Kuzma's mother" to Sukhodrev and explained that he had used it to mean "We're going to show you something you've never seen!"; "We're going to catch up and overtake you and surprise you with something you've never seen!" The phrase became one of the slogans of the Cold War, and the variation, "Kuzka's Mother" became a metaphor for the atomic bomb in the subsequent crises of the Cold War. Cold War (Muskin, Adam. s/d Shevcenko 2020; On this day: Russia in a click s/d).

However, on many occasions, even though he knew that what Khrushchev said made no sense, he had to follow his words. Arriving in New York in 1960 with Gromyko and Khrushchev, he called attention to advanced construction techniques, but Khrushchev rejected this, and Sukhodrev knew he would have to translate Khrushchev's ideas on best building practices, knowing full well that they were absurd, but he had to follow the norms of his profession (Sukhodrev 1999).

Sukhodrev had a remarkable gift for mimicry and adapted his interpreting style to suit the audience. Depending on his listener, he could rapidly switch from perfectly accented British English to American English[49]. And on his visits to India he would unconsciously begin to produce elements of the Indian accent (Spotlight RT 2009).

3. "We will bury you"

Despite the above accolades, at a key moment Sukhodrev was heavily criticized. On 18 November 1956, Nikita Khrushchev uttered the famous words that, according to journalist Danny Kane, set back relations between the Soviet Union and the United States for a decade (Kane 2020: Internet). At the Polish embassy in Moscow, to a room full of Western diplomats, he

---

[49] Sukhodrev's American accent can be heard translating Brezhnev at
https://www.youtube.com/watch?v=XjKGYBzlp58
(White House Tapes Conversation) and Khrushchev at
https://www.youtube.com/watch?v=tHNWpgSI_p4&feature=youtu.be
(Velikaia Rossia).

proclaimed: "Мы вас похороним!" translated by Sukhodrev as "We will bury you".

There is much speculation about what Khrushchev meant by "My vas pokhoronim". However, Sukhodrev said he had translated Khrushchev's speech literally and was making simultaneous interpretation, which didn't give him time to think or look for alternatives. "My vas pokhoronim" can be interpreted in different ways, such as: "We will live to see him buried"; "We will be present at your funeral"; "We will live longer than you"; and "We'll outlast you". All of these, while provocative statements, lack the overt threat of "We will bury you".

By taking the full quote in context, his interpretation again loses some of the threatening tone: "Whether you like it or not, we are on the right side of history. We will bury you". In this context, the phrase refers more to ideology and the construction of history than to war. In 1963, Khrushchev recalled his speech saying, "I once said, 'We're going to bury you,' and I got in trouble for that. Of course, we will not bury them with a shovel. Their own working class will bury them" (Kane 2020).

4. The Photographs

Photograph 1: "He's got two glasses!"

In *Photograph 1*, from 19 June 1973, at a meeting between the United States and the USSR to try to reach a permanent agreement for the limitation of strategic arms (SALT – Strategic Arms Limitation Talks), we see a toast between Brezhnev and Nixon, who are apparently getting along well. Here Brezhnev holds two glasses of champagne, Nixon none, pointing at Brezhnev with a mock-mean grimace, as if he were jokingly snitching on him and telling a classmate Brezhnev was drinking too much. And indeed Brezhnev did have a drinking problem, as confirmed by Andrei Gromyko (L. A. Times Archives 1989). In his later years, he was dependent on sleeping pills and alcohol. Mark Lawrence Schrad (2013), following the Memoirs of Anatoly Dobrynin, Soviet ambassador to the United States, reports that, after the meeting in Washington D.C., Brezhnev stayed at Nixon's ranch in California. After a formal dinner Brezhnev continued drinking whisky and late that night was drunk and then bumped into Nixon in the hallway. Dobrynin, who didn't drink and had to act as interpreter, was mortified by the fact that Brezhnev seemed on the verge of revealing the secrets of intrigues in the Kremlin and kept complaining about his Politburo comrades. Ultimately, Dobrynin and Nixon finally managed to take Brezhnev to bed.

In *My Language is My Friend* Sukhodrev comments on the large amount of alcohol consumed at political receptions and diplomatic visits. It seems that there was a certain obligation for Soviet politicians to drink a lot. Khrushchev got around this problem of always having to drink on official occasions by pouring himself water from a bottle of Stolichnaya vodka (Sukhodrev 1999) or using a special goblet he received as a gift from Jane Thompson, wife of the US ambassador, which gave to impression that it was always full of vodka (Sukhodrev: "Visiting Dmitry Gordon" 2/3). Likewise, Brezhnev was also poured water our of a Stolichnaya bottle (Sukhodrev 1999:149).

Andrei Rogatchevski (2019) comments that, unlike a court or congress interpreter, the high-level political interpreter spends a lot of time with the same person, often working long hours, even attending social events after work and frequently developing an important bond with the politician, and even, as in Brezhnev's case, the politician may develop a certain dependence on the interpreter. At Brezhnev's meeting with Jimmy Carter in Vienna in June 1979 for the SALT (Strategic Arms Limitation Treaty) talks, Brezhnev gave Carter the greeting common among friends in the USSR, a kiss on either side of the face. When Sukhodrev was translating Carter's words for Brezhnev, he stopped him and asked if he had done the right thing. Sukhodrev stifled his laughter and told him that everything seemed normal, and it was the right thing to do at the time, thus calming Brezhnev (Sukhodrev 1999).

Sometimes it seems that Sukhodrev has something in common with the guide/interpreter who, accustomed to international meetings, has to guide the politician through the jungle of protocols and etiquette. He mentions that he had to buy clothes for Soviet leaders at the request of Foreign Minister Gromyko; advise Gromyko on which appetizer to choose; and explain to Brezhnev what an appetizer was and which spare parts of the Lincoln Continental limousine he had received as a gift from the Lincoln-Ford company he might need in the future (Sukhodrev 1999; Rogatchevski 2019).

Jesus Baigorri-Jalón and Maria Manuela Fernández-Sánchez discuss the role of interpreters as caregivers for important figures in China: "The influence of Nancy Tang and Wang Hairong became increasing great over time, particularly at the end of Mao's life". When the dying Helmsman received foreign dignitaries he could hardly speak, and his interpreters had to decide what he wanted to say to the visitors (Holdridge 1997 in Baigorri-Jalón and Fernández-Sánchez 2010). The same authors also mention that the interpreter Ji Chaozhu played the role of one of the bodyguards at the Congress of Geneva in 1961 when there was the possibility of an attempt on the life of Premier Zhou (Baigorri-Jalón and Fernández-Sánchez (2010).

Nixon also had his drinking problems. Merrill Fabry points to Nixon's difficulty with alcohol: "Bob Woodward and Carl Bernstein wrote a book about Nixon's last days in the White House in which they note that the President's inability to handle more than one drink was well known to his intimates" (Fabry 1982). So in our photograph, Nixon, who already looks a little tipsy and quite relaxed after a glass of champagne, banters with Brezhnev. This scenario could end badly, with an embarrassing situation at this very delicate moment for the future of the planet! So the one who has to handle the situation is the interpreter. Used to guiding and caring for Brezhnev, he now finds Nixon quite merry and so has another charge to look after. He's in the middle, quickly translating the two lads' comments and jokes, controlling the situation and making sure it doesn't end badly.

Indeed, the concept of the interpreter as a guide/caregiver runs through this book. In Chapter 2 we saw Julius Meyer trying to introduce the Pawnee Indians to capitalism; the SPI interpreters in Chapter 3 were all guide-scouts-interpreters; Chang in Chapter 5 was John Thomson's guide-interpreter; and in Chapter 6 John Brown was Captain Thomas Whiffen's guide-interpreter in the Putumayo.

The confidence Nixon had in Sukhodrev while distrusting official interpreters such as Harry Obst, whom we saw in Chapter 7, seems extraordinary. Baigorri-Jalón and Fernández-Sánchez, following William Burr, describe the distrust of the State Department on the part of Nixon and Kissinger, who shared the same

> preoccupation with secrecy, their exaltation of presidential authority and control, their contempt for bureaucracy and Congress, and their deeply manipulative conception of politics in United States' international relations (Burr 1998, in Baigorri-Jalón and Fernández-Sánchez 2010).

And there are indications that Nixon was also somewhat dependent on Sukhodrev. Before giving a speech on live television on 28 May 1972, Nixon asked Sukhodrev to read the Russian version of the speech backstage, with the same emotional charge as he himself would give it. Sukhodrev was in a difficult position because one might think that he, the interpreter, shared the ideas of US policy (Sukhodrev 1999). However, as William Safire wrote in the New York Times, Sukhodrev's performance was excellent:

> Soviet viewers who saw Sukhodrev, the best in the business at the top of his form – not drily translating, but dramatically driving home Nixon's mood and message. Obviously, no one told him not to do his professional best (Safire 1972:13, in Baigorri-Jalón and Fernández-Sánchez 2010).

After he interpreted Gorbachev's interview for NBC in 1987, he received a congratulatory telegram from Nixon, which reminded him of the wonderful work he had done in the 1972, 1973, and 1974 meetings with Brezhnev (Sukhodrev 1999:147).

And Kissinger also took advantage of Sukhodrev, asking him to dictate his notes on the first meeting between Nixon and Brezhnev in 1972 to his secretary (Kissinger 1979, in Baigorri-Jalón and Fernández-Sánchez 2010).

Sukhodrev also convinced the State Department to use the term "Soviet people" rather than "people living in the USSR" on the Voice of America radio station, which broadcast US news around the world (Sukhodrev 1999:96).

We can see *Photograph 1* as a triptych, a very common form of religious painting, usually with God or Jesus Christ in the centre and saints, faithful, or admirers at the sides. In the central panel of "The Elevation of the Cross" (1610-1611), by Peter Paul Rubens, the cross, with Christ nailed to it, is being lifted up. "The Triptych of the Annunciation" (1425-8), by the Dutch Robert Campion, shows the scene of the Annunciation in the central panel, the donors (who commissioned the painting) in the left-hand panel, and Joseph, with his carpenter's tools, in the right-hand panel. Triptych icons are also very common in Russian religious art.

Thinking of our photograph as a triptych, the supporting actors are clearly Brezhnev and Nixon, with Sukhodrev occupying the main position, and Brezhnev and Nixon showing their reverence to Sukhodrev. There are certain similarities to Julius Meyer's photos in Chapter 2, but here the relationship is much more subtle. Sukhodrev comments that "It's an almost mystical feeling that you are bringing people together […] people who otherwise would never be able to communicate" (Schudel 2014), and here we see evidence of this sublime ability. However, he could have added something like "and look after them when they are together!"

Several videos show Sukhodrev as Brezhnev's minder: "President Nixon Welcomes Leonid Brezhnev to the United States" (President Nixon Welcomes Leonid Brezhnev to the United States. 1973) and "Nixon Welcomes Brezhnev at White House" (SYND 18 6 73. 1973).

Sukhodrev was also Khrushchev's personal interpreter and can be seen in a large number of videos with him, always behind his shoulder, as is the case in "A Cool Welcome In NYC – A Clip From "Cold War Roadshow"; and "Khrushchev does America" (1959), from Khrushchev's visit to the United States in 1959. Unfortunately, in this documentary we do not hear Sukhodrev's voice, with his American accent, but a narration with a strong Russian accent, probably to give impression of greater estrangement and proximity to the way Khrushchev would have spoken English.

In "Nikita Khrushchev on Face the Nation" (Face the Nation 1957) we see Sukhodrev alongside Khrushchev translating into Russian but not into English.

Photograph 2: Toast!

Photograph 2 was taken shortly before *Photograph 1*. It seems that Nixon was not very well informed about the Russian toasting tradition, but he appears to be enjoying this moment of relaxation, and, with his weak resistance, perhaps the drink is already starting to have some effect. Sukhodrev is half smiling, wondering how the situation will work out. He may have to take care of Nixon as well as his boss. Andrei Gromyko, who always had a reputation for being very responsible and dedicated and who has no doubt witnessed several instances where Brezhnev got drunk, looks worried and wipes his mouth with his handkerchief. William F. Rogers (1913-2001), Secretary of State from 1969 to 1973, but who always had a secondary role in relation to Henry Kissinger (1923-2023), the National Security Advisor, seems happy, despite not having received a glass of champagne.

In *My Language is My Friend* Sukhodrev writes: "The great challenge of my profession: to be present, but invisible"[50] (Sukhodrev 1999). However, the photos we've seen of Sukhodrev belie that claim. Sukhodrev is always there, in the centre of the picture, wise, aware, calm, in control, always well dressed and looking relaxed. Sukhodrev was even in a prominent position

---

[50] "В этом высший пилотаж моей профессии: стать как бы невидимым, но присутствующим".

on the cover of *LIFE*, probably the only interpreter ever to appear on the magazine's cover.

Photograph 3: The Glassboro Cupula from LIFE cover, June 30, 1967

The Glassboro Summit Conference was held from 23 to 25 June 1967 between the heads of government of the US and USSR, President Lyndon B. Johnson and Premier Alexei Kosygin, in Glassboro, New Jersey, to discuss the situation in Middle East after the Six Day War between Israel and Egypt, and the Vietnam War. The result was positive, improving the atmosphere between the USA and the USSR at the height of the Cold War.

As we saw in Chapter 7, *LIFE* was always keen to make a positive impression of and for the United States, and this can be seen in Ben Martin's cover photo. President Lyndon Johnson is smiling, fatherly and good-humoured, while Premier Andrei Kosygin looks sullen, declining to look at the American President, apparently refusing contact with Johnson, the representative of the West. The message is clear: "We, the West, want contact with you, but you want to live in a world apart, a sad world where nobody smiles".

As always, Sukhodrev stands out in the photos. The caption reads: "Kosygin and Johnson with interpreter in Glassboro". As we saw in the photos about the visit of the Chinese delegation to Issy-les-Moulineaux in May 1910, the press had little interest in the figure of the interpreter, choosing a photo that does not show Vissière, and, here, as in the edition of *LIFE* on the attack on Nixon's car in Caracas, not mentioning Sukhodrev by name. An interesting point is that there is no mention as to whether the interpreter is American or Russian. It is possible that if they had listened to his American accent, reporters might have thought that Sukhodrev was American rather than Russian. If the figures of Johnson and Kosygin are static, Sukhodrev is speaking, gesturing and facing Kosygin. It looks as if he is translating something Johnson just said. In Chapter 4 we saw the important symbolism of the hand in photographs, and here Sukhodrev's hand gesture can be read as something along the lines of: "The sympathetic and smiling President Johnson is offering you something, but you are turning your back and don't want to know about the offer", reinforcing the idea that *LIFE* chose a photo that gave a positive impression of Johnson and a negative one of Kosygin, dull and shut in his world, indifferent to the interpreter's requests. Indeed, *My Language is My Friend* includes a photo of a smiling Kosygin (1999:74). However, the preference was for the apparently dour and mirthless Soviet.

That day Sukhodrev had a major pain problem after a tooth extraction in New York the day before, and in the photo his left cheek looks quite swollen. However, this does not seem to affect his ability to interpret (Sukhodrev 1999).

Photograph 4: The English Boy

Photograph 5: Uncle Jack and Auntie Mary

In *Photograph 5*, dated 2 February 1967, just a few months before the Glassboro summit, we see Walter and Catherine Holloway, Uncle Jack nd Aunt Mary, looking at *Photograph 4* of Sukhodrev in London, next to the flat in Hampstead where the boy and his mother lived. Sukhodrev is visiting London with Kosygin. On this visit he wanted to meet the Holloways, but, now retired, they were living on the Isle of Wight and could only be contacted by phone. In the photo Sukhodrev looks at a photograph of himself as a boy, taken and developed by Uncle Jack in his dark room. The text that accompanies the photo on "Agefotostock" demonstrates the great affection and affection that Uncle Jack had for young Viktor:

> Pointing to a Tommy-Gun he was carrying in the picture of himself, he said "I was the envy of all boys in the road with that gun. Uncle Jack made it for me from a piece of gas piping, a tin can and part of an old chair back. He even put a little motor inside which gave it a very authentic 'click-click-clicking sound. I took the gun back to Russia with me. Now is belongs to my 11 year old son, Sergei. The tin hat I'm wearing was the one Uncle Jack wore in the First World War. He stuck a Red Star on the front of it – and I felt very proud" (Agefotostock Photo).

Let's try to untangle the images of the *mise en abîme* of this photo. The Holloways, both in their 70s and retired, living in Ryde on the Isle of Wight, look at a man, now in a very important position in a distant, powerful, and inaccessible country, on the other side of the Iron Curtain, who was a substitute son for them and whom they, who had no children, loved. Uncle Jack proudly points to Viktor: "This man, so well-presented, so handsome, so successful, was my great friend, Viktor, almost my son, and when he could he accompanied me on my rounds to deliver the post".

In the second photo Viktor is next to the apartment where he lived between the ages of six and twelve, from 1939 to 1945, and where his destiny was decided. Despite the Second World War, which lasted throughout Viktor's stay in London, he writes with great nostalgia about his time in London. He looks at the third picture, a picture of himself with the toy tommy gun made by Uncle Jack. The boy is confident, proud of his toys and, with the Soviet star on his helmet, ready to defend the honour of the Motherland. The toy machine gun becomes a symbol of his connection to the Holloways and is passed down to the next generation, his son Sergei. And the helmet, from the First World War, was also Uncle Jack's.

And so we end our chapter on Viktor Sukhodrev, the interpreter who, during the Cold War, was always there, the man in the middle, "the voice in the English language of all Soviet leaders from Nikita Khrushchev to Mikhail Gorbachev" (Mydans 2005).

AFTERWORD

This book, originally written in Portuguese as *Fotografia de Intérpretes: Em Busca de Vidas Perdidas* and published by Editora Lexikos, São Paulo, at the end of 2022, involved the selection and discarding of subjects for chapters. I began the study with the chapters on Vissière and Julius Meyer, Chapters 1 and 2, and then Chapter 4, on Brazilian indigenous translators, and Chapter 3, on the SPI. I had known about Vernon Walters and began to read his autobiography and works about him and the 1964 Brazilian military coup, dealt with in Chapter 6. I then came across the photo of John Thomson of Chang and the Young Well-dressed Indigenous Interpreter, subjects of Chapter 5. And then the book was finalised by Chapter 8, on Viktor Sukhodrev, and Chapter 6, on John Brown.

I also considered other chapters: on Robert Hart (1835-1911), the Northern Irish consular interpreter in China (1854-59), who spent most of his career as the head of the Chinese Maritime Customs Bureau, which collected Chinese tariffs on foreign imports, by 1895 employing more than 700 Westerners and 3,500 Chinese The Bureau also charted the China coast, managed government port facilities, and supervised the lighting of coastal and inland waterways, and was enormously important for the finances of the Empire.

Other possibilities were to write on photographs the interpreters of politicians who have been of major importance in the last hundred years or so: Stalin, Mao Tsé-Tung, Putin, Trump, for example. Much has been documented in works such as Valentim's Berezhkov's *At Stalin's Side* (1994), Pavel Palazchenko's *My years with Gorbachev and Shevardnadze: the memoir of a Soviet Interpreter* (1997), and Li Chaozhu's *The Man on Mao's Right: From Harvard Yard to Tiananmen Square. My Life Inside China's Foreign Ministry* (2008). However, I failed to find photographs which were interesting to analyze.

Other possibilities came up. The Mayor of New York from 1934 to 1946, Fiorello H. La Guardia (1882-1947) had worked as an immigration interpreter at Ellis Island from 1907 to 1910 in Italian, German, Yiddish and Croatian. But, as in the previous cases, I found no photographs of La Guardia actually interpreting.

In a completely different area, the outspoken football manager José Mourinho (1963-), ex-manager of Benfica, União de Leiria, Porto, Chelsea, Inter Milan, Real Madrid, Manchester United, Tottenham Hotspur, and Roma, and, as I write, recently appointed manager at the Turkish club, Fenerbahçe. He began his career as an interpreter for Sir Bobby Robson at Sporting CP and Porto in Portugal, then as Robson's interpreter and assistant at Barcelona. Mourinho would characteristically add some of his own information after translating Robson's words, somewhat similar to the way in which we saw Megaron translating Raoni in Chapter 4. Mourinho can be seen interpreting Robson on the Internet.

Hopefully, these interpreters will be dealt with in further studies.

Bibliography

Agefotostock Photo - Feb. 02, 1967 - Russia's top interpreter looks at a picture which takes him back 25 years. https://www.agefotostock.com/age/en/Stock-Images/Rights-Managed/ZUK-19670202-baf-k09-052. Accessed on 26 September 2020.

Aird, Michael. 2003. "Growing up with Aborigines", in Pinney, Christopher and Nicolas Peterson (eds.). 2003. *Photography's Other Histories.* Durham (US) and London: Duke University Press. 23-39.

All Things Considered. 2014. "NPR Interpreter Viktor Sukhodrev, Fulcrum Of The Cold War, Dies At 81". May 19, 2014. https://www.npr.org/2014/05/19/313996727/interpreter-viktor-sukhodrev-fulcrum-of-the-cold-war-dies-at-81. Accessed on 4 August 2024.

Amazônia Real: "Os Bolsonaro, segundo o cacique Megaron Txucarramãe". 8 May 2018. https://amazoniareal.com.br/os-bolsonaro-segundo-o-cacique-megaron-txucarramae/. Accessed on 26 September 2024.

American tribes.com. Julius Meyer. http://amertribes.proboards.com/thread/2930/julius-meyer. Accessed on 4 August 2024.

Andrews, Evan. 2020. "The History of the Handshake: The ritual gesture has existed since ancient times – but its use as an everyday greeting is a more recent phenomenon. history.com. March 16, 2020. https://www.history.com/news/what-is-the-origin-of-the-handshake. Accessed on 23 September 2020.

Appelbaum, Henry R. "In memoriam, Vernon Walters: Renaissance Man". https://www.cia.gov/library/center-for-the-study-of-intelligence/kent-csi/vol46no1/pdf/v46i1a01p.pdf. Accessed on 13 September 2020.

Apple, R.W. Jr. 1973. "Nixon, Brezhnev Give Peace Vows as Summit Begins: 3¾Hour Meeting; Soviet Aide Describes Session as 'Good, Businesslike'". *The New York Times*, June 19, 1973. https://www.nytimes.com/1973/06/19/archives/nixon-brezhnev-give-peace-vows-as-summit-begins-3-34hour-meeting.html. Accessed on 4 August 2024.

Arisi. Patricia. 2018. "Os Magníficos Korubo: um estúdio de Sebastião Salgado na selva amazônica". 18 January 2018. Amazonia Real.

https://amazoniareal.com.br/os-magníficos-korubo-um-estudio-de-sebastiao-salgado-na-selva-amazonica/. Accessed on 4 August 2024.

Arnheim, Rudolf. (1954/1974). *Art and Visual Perception*. Berkley: University of California Press.

Arnheim, Rudolf. (1988). *The Power of the Center*. Berkley: University of California Press.

Arbus, Diana. https://www.azquotes.com/author/499-Diane_Arbus. Accessed on 4 August 2024.

Arquivo Histórico Ultramarino (AHU). https://digitarq.ahu.arquivos.pt/details?id=1187215. Accessed on 4 August 2024.

Aubin, Jean (1974). "Francisco de Albuquerque, un Juif castillan au service de l'Inde portugaise (1510-1515)". Arquivos do Centro Cultural Português, VII. Lisboa/Paris.

Baigorri-Jalón, Jesús. 2004. *Interpreters at the United Nations: a History*. Translation from Spanish by Anne Barr. Salamanca: Ediciones Universidad Salamanca.

Baigorri-Jalón, Jesús, and María Manuela Fernández-Sánchez. 2010. "Understanding High Level Interpreting in the Cold War: Preliminary Notes", in *Forum*, Vol. 8, No. 2, October 2010. 1-29.

Baigorri-Jalón, Jesús. 2016a. "Interpreters, photography and memory: Rabinovitch's private archive". *The Interpreters' Newsletter* 2016 (21), 1-16.

Baigorri-Jalón, Jesús. 2016b. "The use of photographs as historical sources, a case study: Early simultaneous interpreting at the United Nations". *New Insights in the History of Interpreting*, ed. Kayako Takeda and Jesús Baigorri-Jalón. Amsterdam: John Benjamins.

Bandeira, Luiz Alberto Moniz. 1977. *O Governo João Goulart: As Lutas Sociais no Brasil (1961-1964)*. Rio de Janeiro: Civilização Brasileira.

Barbio, Luciana Alves. 2005. "Identidade e representação: uma análise da sociedade Paresí através do discurso sobre as fotografias da Comissão Rondon". MA dissertation, Postgraduate Programme in Sociology and Anthropology, Universidade Federal do Rio de Janeiro.

Barbosa, Helena Lúcia Silveira, and John Milton. 2020. "Photographs of the interpreters of the Indian Protection Service – SPI (1910-1967). *Tradução em Revista* 28, 2020.1 66-86. https://www.maxwell.vrac.puc-rio.br/48168/48168.PDFXXvmi=.

Barbosa, Helena. 2020. e-mail: 18 September 2020.

Barbosa, Luiz Bueno Horta. 1947. "A pacificação dos índios caingangue paulista". *O problema indígena no Brasil*. Rio de Janeiro: Imprensa Nacional (Comissão Rondon Publicação 88).

Barthes, Roland. 1980/1981. *Camera Lucida: Reflections on Photography Câmara Clara.* Translation by Richard Howard of *La chambre claire: note sur la photographie. New* York: Hill & Wang..

BBC. 2005. "Condor legacy haunts South America", June 8, 2005. http://news.bbc.co.uk/2/hi/americas/3720724.stm. Accessed on 4 August 2024.

Beliel, Ricardo. 2013. "A Frente de Contato dos Korubo". O Índio na Fotografia Brasileira. http://fotografia.povosindigenas.com.br/ricardo-beliel/. Accessed on 4 August 2024.

Benjamin, Walter. 1931. "A short history of photography". *Screen*, Volume 13, Issue 1, Spring 1972. 5-26, https://doi.org/10.1093/screen/13.1.5. Accessed on 7 August 2024.

Berezhkov, Valentim M. 1994. At Stalin's Side. New York: Birch Lane Press.

Berger John 1967/2013. *Understanding a Photograph.* Harmondsworth: Penguin.

Berger, John. 1980. "Understanding a Photograph", in *Classic Essays on Photography*, ed. Alan Trachtenberg. New Haven, Conn.: Leet's Island Books. 291-294.

Berger, John. 1980. "Uses of photography", in *About Looking.* New York: Pantheon.

Bessa Freire, José Ribamar. 2020. e-mail: 28 September 2020.

Boteko Vermelho. 2010. "Garoto-símbolo da reeleição de Lula diz que sua vida mudou para melhor nos últimos anos", 27 December 2010. http://botekovermelho.blogspot.com/2010/12/garoto-simbolo-da-reeleicao-de-lula-diz.html. Accessed on 4 August 2024.

Bouchon, Geneviève. 1985. *L'interprète portugais en Inde au début du XVIe siècle. Simpósio Interdisciplinar de Estudos Portugueses, As Dimensões da Alteridade nas Culturas de Língua Portuguesa - o Outro, II.* Lisboa: Universidade Nova de Lisboa.

Brañas, Manuel Martín. 2013. "Silvino Santos: Arte y Propaganda en la Época del Caucho" in *Álbum de Fotografías: Viaje del Comisión al Rio Putumayo y sus Afluentes, Agosto a Octubre de 2012.* Iquitos: Programa de cooperación Hispano Peruano. 15-19.

Brassai. https://www.azquotes.com/author/20616-Brassai. Accessed on 4 August 2024.

Brecht, Bertolt. 1976. "Portrayal of Past and Present in One", from Bertolt Brecht: *Plays, Poetry and Prose, Poems 1913-1956.* ed. John Willett and Ralph Manheim with the co-operation of Erich Fried. London: Eyre Methuen. 307.

British Pathé News. 1958. Tragic Incidents In Venezuela Against Vice President Nixon. https://www.youtube.com/watch?v=EHR1dBTJrRA. Accessed on 4 August 2024.

Bryson, Bill. 2013/2014. *One Summer: America 1927*. London: Black Swan.

Burke, Peter. 2001. *Eyewitnessing: The Uses of Images as Historical Evidence*. Ithaca: Cornell University Press.

Burr, William 1998. *The Kissinger Transcripts. The Top Secret Talks with Beijing and Moscow*. New York: The New York.

Cabrera Becerra, Gabriel. 2018. "La presencia antillana en la Amazonia: los negros barbadenses en la explotación del caucho y sus imágenes", in Memorias: Revista Digital de Historia y Arqueología desde el Caribe, núm. 36, 2018. http://www.scielo.org.co/scielo.php?script=sci_arttext&pid=S1794-88 862018000300057&lng=en&nrm=iso&tlng=es. Accessed on 4 August 2024.

Cadamosto, Luis. 1948. *Viagens de Cadamosto e de Pedro Sintra*. Translation by João Franco Machado. Preface and notes by Damião Peres. Lisboa: Academia Portuguesa de História.

Cain, Abigail. 2017. "How Frederick Douglass Harnessed the Power of Portraiture to Reframe Blackness in America", in *Artsy*. https://www.artsy.net/article/artsy-editorial-frederick-douglass-photographed-american-19th-century. Accessed on 4 August 2024.

Callado, Eduardo, and Artur Silveira da Mota. 1880/2012. Ofício 5, 20 fev. 1880, ahi 271/01/20 "Índice: Contrato de um intérprete em Paris". *Cadernos do CHDD* (Centro de História e Documentação Diplomática). Brasília: Fundação Alexandre de Gusmão (FUNAG). 46-47.

Campos, Paula Cristina F. 2006. "Intérpretes, Topazes & Línguas: Elos de ligação dos portugueses, de Lisboa a Macau (Séculos XV, XVI e XVII)". Universidade de Macau, M.A. dissertation in Portuguese Studies – History. Macau.

Cartier-Bresson, Henri. azquotes.com. https://www.azquotes.com/quote/646056. Accessed on 4 August 2024.

Cartier-Bresson, Henri. PhotoQuotes.com. https://www.photoquotes.com/printableshowquotes.aspx?id=98. Accessed on 4 August 2024.

Carvalho da Cruz, Ariane. 2015. "Sempre Vassalo Fiel de Sua Majestade Fidelíssima": Os autos de vassalagem e as cartas patentes para autoridades locais africanas (Angola, segunda metade do século XVIII). *Cadernos de Estudos Africanos* [Online], 30 | 2015. http://journals.openedition.org/cea/1834. Accessed on 6 August 2024.

Carvalho, F. M. 2010. "Ngolas, sobas, tandalas e macotas: hierarquia e distribição de poer no antigo reino do Ndongo", in Ribeiro, A.; Geabra, A.; Bittencourt, M. (orgs.). *África passado e presente: II encontro de estudos africanos da UFF* [recurso eletrônico] – Niterói: PPGHISTÓRIA-UFF.

Carvalho, F. M. 2013. "Os Homens do rei em Angola: sobas, governadores, capitães mores, séculos XVII e XVIII". PhD in History. Universidade Federal Fluminense, Niterói, 2013.

Carvalho, Joaquim de. 2019. "Lula e as mãos que fazem o Brasil", in DCM, 10 October 2019. https://www.diariodocentrodomundo.com.br/lula-e-as-maos-que-fazem-o-brasil/. Accessed on 28 December 2023.

Casement, Roger. 2016. *Diário da Amazônia de Roger Casement*. Introduction ad selection by Angus Mitchell. Translation by Mariana Bolfarine (coord.), Mail Marques de Azevedo, and Maria Rita Drumond Viana. São Paulo: EDUSP. Journal, 1997:178.

Castanheda, Fernão Lopes de. 1979. *Historia do Descobrimento e Conquista da India pelos Portugueses*. Porto: Lello & Irmão.

Chapparro, M. and Chirif A. 2013. 'Introducción' in *Álbum de Fotografías: Viaje del Comisión al Rio Putumayo y sus Afluentes, Agosto a Octubre de 2012*. Iquitos: Programa de cooperación Hispano Peruano. 9-13.

Chaozhu, Ji. 2008. *The Man on Mao's Right: From Harvard Yard to Tiananmen Square. My Life Inside China's Foreign Ministry*. New York: Random House.

Chaumeil, Jean-Pierre. 2013. "Una Misión Diplomático al Putumayo en 1912", in *Álbum de Fotografías: Viaje del Comisión al Rio Putumayo y sus Afluentes, Agosto a Octubre de 1912*. Iquitos: Programa de cooperación Hispano Peruano. 21-23.

Chaumeil Jean-Pierre. 2009, "Guerra de imágenes en el Putumayo (1902-1920)", in Chirif, Alberto, y Manuel Cornejo Chaparro (eds.), *Imaginario e imágenes de la época del caucho: los sucesos del Putumayo*, CAAAP/IWGIA/UPC, Lima. 37-73.

Chin/Jaw Face Reading: https://www.yourchineseastrology.com/face-reading/forehead/chin/. Accessed on 8 April 2020.

Chirif, Alberto, and Manuel Cornejo Chaparro, eds.. 2009. *Imaginario e imágenes de la época del caucho: Los sucesos del Putumayo*. Lima: Centro Amazónico de Antropología y Aplicación Práctica (CAAAP).

CIA Library. 2012. Biography: Lieutenant General Vernon A. Walters. https://www.cia.gov/library/readingroom/docs/CIA-RDP99-00418R000100380001-0.pdf. Accessed on 26 September 2020.

Clarke, Graham. 1997. *The Photograph: a Visual and Critical History* Oxford: Oxford University Press.

Claudel, Paul. 1957. *Cinq Grandes Odes, dans Oeuvre Poétique* Paris: Pléiade.

Coalson, Robert. 2013. "A Surprisingly Candid Chat Between Nixon and Brezhnev: Despite Cold War tensions, newly released tapes show the two found plenty of room for jokes in a 1973 meeting", in *The Atlantic*, August 26, 2013.
https://www.theatlantic.com/international/archive/2013/08/a-surprisingly-candid-chat-between-nixon-and-brezhnev/278981/.
Accessed on 5 August 2024.

Contreiras, Hélio. 2001. "O general Vernon Walters defende o ex-ditador chileno, ataca Fidel Castro e fala da participação dos EUA nos preparativos do golpe militar de 1964". *Isto É*. Edição 04/04/2001 - n° 1644.

Correa, Tom. 2013. "The American Cowboy Chronicles: The Pawnee Massacre - Sundown of the Pawnee Indians", December 5, 2013.
http://www.americancowboychronicles.com/2013/12/the-pawnee-massacre-sundown-of-pawnee.html. Accessed on 17 June 2020.

Correia, Gaspar. 1922. *Lendas da India*. Coimbra: Na Imprensa da Universidade.

Couto, Dejanirah. 2003. "The Role of Interpreters, or Linguas, in the Portuguese Empire During the 16th Century". e-JPH, Vol. 1, number 2, Winter 2003 https://digitalis-dsp.uc.pt/bitstream/10316.2/25479/1/EJP H1_2_artigo2.pdf?ln=pt-pt. Accessed on 26 September 2020. Em português. 2011. "O papel dos intérpretes ou línguas no império português do século XVI", in *História do Ensino de Línguas no Brasil* (*HELB*), Ano 5 - No. 5 - 1/2011.
http://www.helb.org.br/index.php/revista-helb/ano-5-no-5-12011/189-o-papel-dos-interpretes-ou-linguas-no-imperio-portugues-do-seculo-xvi-. Accessed on 5 August 2024.

Das Übersetzerportal UEPO.de. 2014. "Viktor Sukhodrev gestorben – Dolmetscher auf höchster Ebene von Chruschtschow bis Gorbatschow".
https://uepo.de/2014/05/18/viktor-sukhodrev-gestorben-dolmetscher-auf-hoechster-ebene-von-chruschtschow-bis-gorbatschow/.   Accessed on 7 August 2024.

Diálogos do Sul. 2017. "Povos Indígenas: Megaron, a Avó do Mundo e a Convenção 169 da OIT", 20 April 2017.
https://dialogosdosul.operamundi.uol.com.br/povos-indigenas/52392/povos-indigenas-megaron-a-avo-do-mundo-e-a-convencao-169-da-oit.
Accessed on 7 August 2024.

Disney, Anthony. n/d. "Smugglers and Smuggling in the Western half of the Estado da India in the late sixteenth and early seventeenth centuries", in *Indica*, 49, Bombay: Heras Institute.

Dubois, Philippe. 1990/2019. *O Ato Fotográfo e outros ensaios*. Translation by Marina Appenzeller of *L'acte photographique et autres essais*. Campinas: Papirus.

Dutilleux, Jean-Pierre, and Luiz Carlos Saldanha, dir. 1978. *Raoni* (também conhecido como *Raoni: The Fight for the Amazon*).

Dyer, Geoff. 2007. *The Ongoing Moment*. New York: Vintage.

Echeverri, Juan Álvaro. 2014. "La suerte de Robuchon", in eds. Claudia Steiner Sampedro, Carlos Páramo Bonilla y Roberto Pineda Camacho, in *El paraíso del diablo: Roger Casement y el informe de Putomayo un siglo después*. Bogotá: Editorial Universidad de los Andes. 233-251.

Edwards, Elizabeth. 2001. *Raw Histories*. Oxford: Berg.

Edwards, Elizabeth. 2014. "Interpreting photographs: Some thoughts on method", in *Framing the Interpreter: Towards a visual perspective*, ed. Anxo Fernández-Ocampo and Michaela Wolf. London: Routledge. 19-26.

Edwards, Elizabeth, and Janice Hart (eds.). 2004. *Photographs Objects Histories: On the materiality of images*. Routledge: New York and London.

Egmond, Florike, and Peter Mason. 1999. "A horse called Belisarius", History Workshop Journal 47:243-52. https://disciplesofflight.com/louis-bleriot-inventor-designer-and-daring-pilot/

Espada, Arcadi. 2000. "Ganamos la guerra fría ENTREVISTA A VERNON A. WALTERS - MILITAR" *El País*, 25 August 2000. https://elpais.com/diario/2000/08/25/opinion/967154411_850215.html. Accessed on 13 September 2020.

Esquerda Bem Informada. 2020. "Raoni defende saída de Bolsonaro 'antes que algo muito ruim aconteça'". 25 September 2020. https://vermelho.org.br/2019/09/25/raoni-defende-saida-de-bolsonaro-antes-que-algo-muito-ruim-aconteca/. Accessed on 5 August 2024.

Executive Intelligence Review, March 25, 2005, "Nazis, Operation Condor, and Bush's Privatization Plan".

Fabry, Merrill. 2016. "Now You Know: What Happens If the President Gets Drunk?", in TIME HISTORY, June 2, 2016. https://time.com/4342019/drunk-president-history/. Accessed on 5 August 2024.

Face the Nation. 1957. Nikita Khrushchev on Face the Nation on June 2, 1957. 60m.

https://www.youtube.com/watch?v=XAG6D73gttA&t=180s. Accessed on 5 August 2024.

Faris, Jamis. 2003. "Navajo and Photography", in Pinney, Christopher and Nicolas Peterson (eds.). 2003. *Photography's Other Histories.* Durham (US) and London: Duke University Press. 85-99.

Fernández-Ocampo, Anxo, and Michaela Wolf. 2014. "Framing the Interpreter", in *Framing the Interpreter: Towards a visual perspective,* ed. Anxo Fernández-Ocampo and Michaela Wolf. London: Routledge. 1-16.

Flores, Jorge Manuel. 1993. "The 'Jurubaças' of Macau, a Frontier Group: the Case of Simão Coelho (1620's)". International Colloquium on Portuguese Discoveries in the Pacific, Santa Barbara: University of California.

Flusser, Vilém. 1983/2000. *Towards a Philosophy of Photography.* Translation by Anthony Mathews of *Für eine Philosophie der Fotografie.* London: Reaktion.

Freire, Carlos Augusto da Rocha. 2005. "Sagas Sertanistas: práticas e representações do campo indigenista no século XX". PhD thesis, Programa de Pós-graduação em Antropologia Social, Museu Nacional, da Universidade Federal do Rio de Janeiro - UFRJ.

Freire, Carlos Augusto da Rocha, and Milton Guran. 2010. Primeiros Contatos: atrações e pacificações do SPI. Rio de Janeiro: Museu do Índio – FUNAI.

Freire, Carlos Augusto da Rocha (org). 2011. Memória do SPI: textos, imagens e documentos sobre o Serviço de Proteção aos Índios (1910-1967). Rio de Janeiro: Museu do Índio – FUNAI.

Freire, Carlos Augusto da Rocha. 1998. "Em Busca da Sociedade Perdida: O trabalho da memória Xetá". M.A. dissertation. Centro de Filosofia e Ciências Humanas, Universidade Federal de Santa Catarina (UFSC).

Frères Caudron. Wikipédia. https://fr.wikipedia.org/wiki/Fr%C3%A8res_Caudron. Accessed on 27 September 2024.

Gaffron, Mercedes. 1950. "Right and left in pictures". *Art Quarterly*, 1950, Vol. 13. 312-313

Gaspardone, E. 1930. "Nécrologie: Arnold Vissière (1858-1930)" Note biographique. *Bulletin de l'École française d'Extrême-Orient*, Année 1930. 649-653. https://www.persee.fr/doc/befeo_0336-1519_1930_ num_30_1_3229. Accessed on 26 September 2020.

Gaspari, Elio. 1980. "A amnésia do general", in *Revista Veja*, 12 de março de 1980.

Gendler, Carol. "The Jews of Omaha: The first sixty years". 1968. M.A. dissertation, University of Nebraska at Omaha. https://digitalcommons.unomaha.edu/cgi/viewcontent.cgi?article=1586 &context=studentwork. Accessed on 26 September 2020.

Ginzburg, Carlo, 1989. "Clues: Roots of an Evidential Paradigm', in Carlo Ginzburg, *Clues, Myths, and the Historical Method.* Translation by John and Anne C. Tedeschi. Baltimore: Johns Hopkins University Press. 96-125.

Giquel, Prosper. 1864/1985 *A Journal of the Chinese Civil War, 1864.* Translation Steven A Leibo. Honolulu: University of Hawaii Press.

Goffman, Erving. 1959/1990. *The Presentation of Self in Everyday Life.* Harmondsworth: Penguin.

Goffman, Erving. 1961. *Encounters: Two Studies in the Sociology of Interaction.* Indianapolis, IM: Bobbs-Merrill.

Gómez Valderrama, Pedro. *Los Infiernos del Jeraca Brown y otros textos.* Bogotá, Fundación Simon y Lola Buberek, 1984.

Goodman, Jordan; 2009. *The Devil and Mr. Casement: One Man's Battle for Human Rights in South One Man's Battle for Human Rights in South America's Heart of Darkness.* London: Verso.

Goodyear, Frank Henry, 1967. *Red Cloud: photographs of a Lakota chief.* Lincoln: University of Nebraska Press.

Guedes, Caroline Coleti, Heitor dos Santos Rodrigues, and Mateus da Cruz Leal Nunes Vitorino. 2018. "Irregularidades no SPI: Corrupção e Violação dos Direitos Humanos Indígenas, entre 1955 e 1967. Undergraduate Project. Universidade Federal do Paraná.

Hardenburg, Walter. 1912. *The Putumayo: the Devil's paradise -Travels in the Peruvian Amazon Region and an Account of the Atrocities Committed upon the Indians therein.* London: T. Fisher Unwin.

Havik, Philip J. 2012. "Gendering the Black Atlantic: Women's Agency in Coastal Trade Settlements in the Guinea Bissau Region", in *Gendering Communities, Economies, and Social Networks in Atlantic Port Cities, 1500-1800, The Atlantic World,* Volume: 25, eds. Douglas Catterall and Jodi Campbell.

Hawthorne, Nathaniel. 1961. *The House of Seven Gables.* New York: Signet.

Hemming, John. 2007. *Ouro vermelho: a conquista dos índios brasileiros.* Translation by Carlos Eugênio Marcondes de Moura. São Paulo: EDUSP, 2007.

Hill, Paul, and Thomas Cooper. 1979/1992. *Dialogue with Photography.* Manchester: Cornerhouse Publications.

Hirsch, Marianne, and Leo Spitzer. 2020. *School Photos in Liquid Time*. Seattle: University of Washington Press. ibooks.

Hockley, Alan. 2010. John Thomson's China – 1: *Illustrations of China and its People*. Photo Albums (1873-1874). https://visualizingcultures.mit.edu/john_thomson_china_01/ct_essay01 .html.

Hoddenbach, Corey. "Louis Bleriot: Inventor, Designer and Daring Pilot", Disciples of Flight. https://disciplesofflight.com/louis-bleriot-inventor-designer-and-daring-pilot/. Accessed on 26 September 2024.

Holdridge, John. H. 1997. *Crossing the Divide*. New York: Rowan & Littlefield.

Hong Kong and Shanghai Tours. "Meeting with Arnold Vissière's descendant". http://shanghaitours.canalblog.com/archives/2017/05/06/35252518.htm l. Accessed on 5 August 2024.

Iglauer, Edith. 1947. "Housekeeping for the family of nations", in *Harper's Magazine*, April 1947:296.

"Índio Caripuna intérprete de Cândido Rondon durante expedições no Amapá". https://www.reddit.com/r/brasil/comments/85fnuv/%C3%ADndio_cari puna_int%C3%A9rprete_de_c%C3%A2ndido_rondon/. Accessed on 5 August 2024.

Ito, A. I. 2016. "Uma 'tão pesada cruz': o governo da Angola portuguesa nos séculos XVI e XVII na perspectiva de Fernão de Sousa (1624-1630)". Dissertação de Mestrado em História Social. Universidade de São Paulo, São Paulo, 2016.

Izarra, Laura P. Z., and Mariana Bolfarine (orgs.). 2016. *Diário da Amazônia de Roger Casement*. Translation by Mariana Bolfarine (coord.), Mail Marques de Azevedo, and Maria Rita Drumond Viana, de *The Amazon Journal of Roger Casement* (1995), ed. Angus Mitchell. São Paulo: EDUSP.

Jackson, Harold. 2002. "Lieutenant General Vernon Walters: "From polyglot soldier to negotiator, presidential aide, deputy director of the CIA, and ambassador to the UN". *The Guardian*, 18 February 2002. https://www.theguardian.com/news/2002/feb/18/guardianobituaries.har oldjackson. Accessed on 5 August 2024.

Johnson, Brooks. 1995. *Photography II. From The Chrysler Museum Collection. 70 Photographers On Their Art*. Norfolk, Virginia: The Chrysler Museum.

Johnson, H. 2002. "Barbadian Migrants in the Putumayo District of the Amazon 1904-1911", in Chamberlain, M. (ed.) *Caribbean Migration: Globalised Identities*. London: Routledge.

John Thomson (Photographer). Wikipedia. https://en.wikipedia.org/wiki/John_Thomson_(photographer). Accessed on 26 September 2020.

Kane, Danny. 2020. Exploring History. "We Will Bury You". – How A Mistranslation Almost Started WW3: And the story of the man behind those fateful words. 14 July 2020. https://medium.com/exploring-history/we-will-bury-you-how-a-mistranslation-almost-started-ww3-4a285162e2b9. Accessed on 5 August 2024.

kenrahn.com. Accessed on 5 August 2024.

Kissinger, Henry. 1979. *White House Years.* Boston: Little, Brown and Company.

Koffman, David S. 2012. "Jews, American Indian Curios and the Westward Expansion of Capitalism", 168-186, in *Chosen capital: the Jewish encounter with American capitalism,* ed. Rebecca Kobrin. Imprint New Brunswick, N.J.: Rutgers University Press.

Kokoszycka, Monika. 2014. "'The Man in the Middle': Viktor Sukhodrev'. Interpreter Soapbox. https://interpretersoapbox.com/conference-interpreting-russians-viktor-sukhodrev-dies/. Accessed on 26 September 2024.

Kolpan, Gerald. 2012. "Blazing Saddles It Wasn't". My Jewish Learning, May 4, 2012. https://www.myjewishlearning.com/members-of-the-scribe/blazing-saddles-it-wasnt/. Accessed on 5 August 2024.

Kolpan, Gerald. 2012. *Magic Words.* New York: Pegasus.

Korchilov, Igor. 1997. *Translating History: Thirty Years on the Front Lines of Diplomacy with a Top Russian Interpreter.* New York: Scribner.

Kracauer, Siegfried. 1980. "Photography", in *Classic Essays on Photography*, ed. Alan Trachtenberg. New Haven, Conn.: Leet's Island Books. 245-268.

Kracauer, Siegfried. 1969. *History: the Last Thing before the Last* (completed and ed. P. O. Kristeller), Princeton, NJ: Marcus Wiener Publications.

Kress, Gunter, and Theo van Leeuwen. 2006 (2nd ed.) Reading Images: The Grammar of Visual Design. London: Routledge.

Kuhn, Annette. 1995/2002. *Family Secrets: Acts of Memory and Imagination.* London and New York: Verso.

Kuhn, Annette. 2003. "Remembrance: The child I never was", in *The Photography Reader,* ed. Liz Wells. London: Routledge. 395-401.

Kuhn, Annette. 2007. "Photography and cultural memory: a methodological exploration". *Visual Studies*, Vol. 22, No. 3, December 2007, 283-292.

Kuhn, Annette, and Kirsten Emiko McAllister. (2006/2008). "Locating Memory: Photographic Acts – An Introduction", in *Locating Memory: Photographic Acts,* eds., Annette Kuhn and Kirsten Emiko McAllister. New York: Berghahn Books. 1-17.

Lange, Dorothea. Dorothea Lange: Imagens e palavras. https://fotodoc.com.br/perfis/dorothea-lange-imagens-e-palavras/

Langford, Rachael. 2014. "Framing and Masking: Photographing the interpreter in/of colonial conflict", in *Framing the Interpreter: Towards a visual perspective*, eds. Anxo Fernández-Ocampo and Michaela Wolf. London: Routledge. 39-50.

Leslie, Esther. 2010. "Siegfried Kracauer and Walter Benjamin: Memory from Weimar to Hitler", in Susannah Radstone, Bill Schwarz (eds.). *Memory: Histories, Theories, Debates*. Fordham University Press. 123-135.

L'Écho de Paris. 1910. "En Plein Air – Aéronautique – "La Mission Chinoise à Issy-les-Moulineaux". 17 de mai de 1910. 6.

Langford, Rachael. 2014. "Framing and Masking: Photographing the interpreter in/of colonial conflict", in *Framing the Interpreter: Towards a visual perspective*, eds. Anxo Fernández-Ocampo and Michaela Wolf. London: Routledge. 39-50.

L. A. Times Archives. 1989. "Afghan Slaying Spurred Invasion, He Says: Gromyko Calls Brezhnev a Problem Drinker", April 4, 1989. https://www.latimes.com/archives/la-xpm-1989-04-04-mn-958-story.html.

Le Gaulois. 1910. "La Mission Chinoise à Issy-les-Moulineaux", 17 de mai de 1910. 3.

Le Grand, G. 1910. "La Mission Chinoise assiste à Issy-les-Moulineaux à une manifestation aéronautique". *L'Auto*, 17 de mai de 1910. 1.

Le Journal. 1910. "La Mission Chinoise à Issy-les-Moulineaux", 17 de mai de 1910. 1.

Le Petit Parisien. 1910. "Les Célestes s'Initient à l'Aéronautique: La mission chinoise à Issy-les-Moulineaux", 17 de mai de 1910. 1.

Le Soleil du Dimanche Illustré, 5 de juin de 1910.

Le Temps. 1910. "La Mission Chinoise à Issy-les-Moulineaux", 17 de mai de 1910. 1.

Les Droits des Enfants: Le travail des enfants: Histoire du travail des enfants en France. https://www.droitsenfant.fr/travail_histoire.htm. Accessed on 16 January 2020.

Lienhard, M. 2005. "Milonga: o 'diálogo' entre portugueses e africanos nas guerras do Congo e de Angola (séculos XVI-XVII)", in Lienhard, M., *O mar e o mato: histórias da escravidão*. Luanda: Editorial Kilombelombe.

Lienhard, Martin. 2008. "Milonga. The "Dialogue" between Portuguese and Africans in the Congo and the Angola Wars (Sixteenth and Seventeenth Centuries), in *Africamericas: itineraries, dialogues, and sounds,* ed. Ineke Phaf-Rheinberger and Tiago de Oliveira Pinto, Madrid/Frankfurt: Iberoamericana/Vervuert. 91-123.

LIFE, May 26, 1958. "Hate Running Loose Hits Out at Nixons: Barely escaping death at hands of a Caracas mob, our V.P. and Pat run a gauntlet of spit and stones". 32-38. https://books.google.com.br/books?id=vVMEAAAAMBAJ&printsec= frontcover&dq=LIFE+-+May+26,+1958+google+books&hl=en&sa=X &ved=0ahUKEwj9943ordnpAhWBKLkGHTvcCHk4ChDoAQhJMA U#v=onepage&q&f=false. Accessed on 5 August 2024.

LIFE, November 25, 1965. "The Blunt Reality of the War in Vietnam", de 26 November 1965. https://books.google.com.br/books?id=FEwEAAAAMBAJ&printsec=fr ontcover&source=gbs_ge_summary_r&cad=0#v=onepage&q&f=false. Accessed on 5 August 2024.

LIFE, June 30, 1967. https://books.google.com.br/books?id=zlUEAAAAMBAJ&source=gbs _all_issues_r&cad=1. Accessed on 5 August 2024.

Lino e Silva, Moises, and Huon Wardle. 2017 "John Brown: Freedom and Imposture in the early Twentieth Century Trans-Caribbean", in *Freedom in Practice: Governance, Autonomy and Liberty in the Everyday,* ed. Moises Lino e Silva, and Huon Wardle. London: Routledge. Pre-publication article (2016).

Lobban, Richard A. Jr. 1995. *Cape Verde: Crioulo Colony to Independent Nation.* Boulder, Colo.: Westview Press, ebook.

Los Angeles Times. 1989. "Afghan Slaying Spurred Invasion, He Says: Gromyko Calls Brezhnev a Problem Drinker" April 4, 1989. https://www.latimes.com/archives/la-xpm-1989-04-04-mn-958-story.html. Accessed on 26 September 2020.

Loti, Pierre.1902/1988. *Les derniers jours de Pékin.* Paris: Calmann-Levy.

Loureiro, Rui Manuel (1992). *Cartas dos Cativos de Cantão: Cristóvão Vieira e Vasco Calvo* (1524?). Instituto Cultural de Macau: Macau.

Machado, Maria Fatima Roberto. 1994. "Índios de Rondon. Rondon e as linhas telegráficas na visão dos sobreviventes Waimare e Kaxiniti, grupos Paresi. PhD thesis in Social Anthropology, Universidade Federal do Estado do Rio de Janeiro.

Mansbridge, Crease. 2016. Mountain Life: Live It Up: Save our Souls: Can a Photograph Steal What is No Longer There? https://www.mountainlifemedia.ca/2016/05/save-our-souls/. Accessed on 6 August 2024.

Mason, George Henry. 1801. Punishing an Interpreter" *The Punishments of China: illustrated by twenty-two engravings.* https://en.wikisource.org/wiki/The_punishments_of_China. Accessed on 5 August 2024.

Mayberry, Kate. 2016. "Third Culture Kids: Citizens of everywhere and nowhere". BBC, Worklife, 18th November 2016. https://www.bbc.com/worklife/article/20161117-third-culture-kids-citizens-of-everywhere-and-nowhere. Accessed on 5 August 2024.

Maybury-Lewis, David. *A Sociedade Xavante.* Translation by Aracy Lopes da Silva. Rio de Janeiro: Francisco Alves.

McCurry, Steve: ttps://www.azquotes.com/author/19888-Steve_McCurry. Accessed on 13 September 2020.

McKenna, T. 1993. *True Hallucinations: Being an Account of the Author's Extraordinary Adventures in the Devil's Paradise.* London: Ebury.

Melo Santos, Rita de Cássia. 2022. "Libânio Koluizorocê. Fragmentos da participação indígena na construção nacional", in *Memórias Insurgentes,* v. 1 n. 1 (2022): Revisitando com os indígenas a formação nacional. Revista UFRJ.

Melo Santos, Rita de Cássia. "Biografia: Major Libânio Koluizorocê. Brasis e suas Memórias". https://osbrasisesuasmemorias.com.br/biografia-major-libaniokoluizoroce/. Accessed on 6 August 2024.

Merencio, Fabiana Terhaag. "Artefatos Líticos da Subcoleção Loureiro Fernandes". *Revista Tecnologia e Ambiente,* Dossiê IX Reunião da Sociedade de Arqueologia Brasileira / Regional Sul, v. 21, n. 1, 2015, Criciúma, Santa Catarina, 2015.

Milton, John. 2024. "*Bolseiros, Lançados, Línguas, Jurubaças* and Other Interpreters of Portuguese in Macau and Africa in the Early Modern Period", in *Translation and Transposition in the Early Modern Period,* eds. Karen Bennett and Rogério Miguel Puga. London: Routledge. 221-235.

MiPutumayo. 2013. "John Brown va hacia donde nace el sol". 24 noviembre, 2013. https://miputumayo.com.co/2013/11/24/john-brown-va-hacia-donde-nace-el-sol/. Accessed on 25 September 2020.

Mitchell, Angus (ed.). 2003. *Sir Roger Casement's Heart of Darkness: The 1911 Documents.* Dublin: Irish Manuscripts Commission.

Mitten, Andy. 2019. "Long read: The making of Jose Mourinho – how did he go from 'the translator' to 'the Special One'?" FourFourTwo: Football, News, Features, Statistics. November 20, 2019. https://www.fourfourtwo.com/features/jose-mourinho-manager-barcelona-profile-translator-special-one-tottenham. Accessed on 6 August 2024.

MONOVISIONS - Black & White Photography Magazine. 2016. Biography: City Life / Documentary photographer John Thomson. 10 May 2016. https://monovisions.com/john-thomson-biography-city-life-documentary-photographer/. Accessed on 6 August 2024.

Montgomery, Hugh. 2004. Keynote, in "Vernon A. Walters: Pathfinder of the Intelligence Profession". Conference Proceedings 3 June 2004. Joint Military Intelligence College. https://ni-u.edu/ni_press/pdf/5482.pdf. Accessed on 6 August 2024.

Murawiec, Laurent. 1984. EIR (Executive Intelligence Review). Volume 11, Number 34, September 4, 1984. "Profile: Vernon Walters: Averell Harriman's paper-clip general". https://larouchepub.com/eiw/public/1984/eirv11n34-19840904/eirv11n34-19840904_051-profile_vernon_walters_averell_h.pdf. Accessed on 26 September 2020.

Muskin, Adam. s/d. "Of Russian origin: Kuzka's mother". RT Russopedia. https://russiapedia.rt.com/of-russian-origin/kuzkas-mother/. Accessed on 6 August 2024.

Mydans, Seth. 2005. "'Man in the middle' at U.S. - Soviet Summits", *New York Times,* Sept. 30, 2005. https://www.nytimes.com/2005/09/30/world/europe/man-in-the-middle-at-ussoviet-summits.html. Accessed on 10 September 2020.

Nathan-Ebrard, Isabelle. 2014. *La Chine: une passion française: Archives de la diplomatie française XVIIIe-XXIe siècles/textes et documents réunis par Isabelle Nathan-Ebrard.* Toulouse: Nouvelles Editions Loubatières.

Nicolson, Harold. 1954. *The Evolution of Diplomatic Method.* Oxford: Oxford University Press.

Nixon's Tour. Vice President Nixon Hostile visit to Peru Warm welcome in Ecuador PublicDomainFootage.com 1958. https://www.youtube.com/watch?v=0PxaqGif_Dg. Accessed on 6 August 2024.

Obst, Harry. 2010. *White House Interpreter: The Art of Interpretation.* Bloomington, Indiana: Author House.

Oliveira, Frank Márcio de. 2005. "General Vernon Walters: gosto por subterrâneo", in *Revista Brasileira de Inteligência*. Brasília: Abin, v. 1, n. 1, dez. 2005. 45-51.

Oliveira, Frank Márcio de. 2019. Attaché Extraordinaire: *Vernon A. Walters and Brazil*/Attaché Extraordinaire: *Vernon A. Walters e o Brasil*. 2019. Washington: NDIC (National Intelligence College) Press.

Paine, Sarah C. M.. 2002. *The Sino-Japanese War of 1894-1895: Perceptions, Power, and Primacy*. Cambridge. 301.

Paiva, Maria Manuela Gomes. 2008. "Traduzir em Macau. Ler o outro – para uma história da mediação linguística e cultural". PhD thesis in Portuguese Studies, Speciality Translation Studies, Universidade Aberta Lisboa.

Paiva, Maria Manuela Gomes. 2011. "Mediadores Linguísticos e Culturais: O 1.º Regimento do Língua da Cidade de Macau". Administração interpreter.º 94, vol. XXIV, 2011-4.º, 1263-1271.

Palazchenko, Pavel. 1997. *My years with Gorbachev and Shevardnadze: the Memoir of a Soviet Interpreter*. University Park, Pennsylvania: The Pennsylvania State University Press.

Panofsky, Erwin. 1939. *Studies in Iconology: Humanistic Themes in the Art of the Renaissance*. New York: Oxford University Press.

Panofsky, Erwin. 1951. *Gothic, Architecture and Scholasticism*. New York: Archabbey.

Parker, Phyllis R.. 1977. *1964: O Papel dos Estados Unidos no Golpe de Estado de 31 de Março*. Translation by Carlos Nayfeld of *U.S. Policy Prior to The Brazilian Coup of* 1964. Rio de Janeiro: Civilização Brasileira.

Paul Mus. Wikipédia: https://fr.wikipedia.org/wiki/Paul_Mus. Accessed on 6 August 2024.

Peixoto, Maurício. 2016. "Mostra na Uerj conta história de ocupação indígena no Rio". *O Globo*, 06/04/2016. https://oglobo.globo.com/rio/bairros/mostra-na-uerj-conta-historia-de-ocupacao-indigena-no-rio-19023725. Accessed on 2 December 2023.

Pelliot, Paul. 1930. "Arnold Vissière", *T'oung Pao*, 27(1), 407-420. https://doi.org/10.1163/156853230X00205. Accessed on 26 September 2020.

Pelliot, Paul. 1948. *Le Hoja Et Le Sayyid Husain de L'histoire Des Ming*. Leiden: Brill.

Peres, Aline. 2007. "Idioma xetá quase morre com Tuca. Índio era o que melhor dominava a língua". *Gazeta do Povo*, 12 June 2007. https://www.gazetadopovo.com.br/vida-e-cidadania/idioma-xeta-quase-morre-com-tuca-aicylj3tdrd2oxr759fgijxou/. Accessed on 6 August 2024.

Peruvian Amazon Co.; 1910. *Correspondence between the Colonial Office and the Foreign Office January to March 1910*. Kew Gardens: National Archives.

Pineda, R. 2000. *Holocausto en el Amazonas. Una historia social de la Casa Arana*. Bogotá: Planeta Colombiana Editorial.

Pinney, Christopher. 1992. "The lexical spaces of eye-spy", in Peter I. Crawford and David Turton (eds.), *Film as Ethnography*. Manchester: Manchester University Press.

Pinney, Christopher. 2003. "Introduction "How the Other Half…"", in Pinney, Christopher and Nicolas Peterson (eds.). *Photography's Other Histories*. Durham (US) and London: Duke University Press. 1-14.

Pinney, Christopher and Nicolas Peterson (eds.). 2003. *Photography's Other Histories*. Durham (US) and London: Duke University Press.

Plotz, David. 2013. "The Greatest Magazine Ever Published: What I learned reading all of *LIFE* magazine from the summer of 1945". Slate. December 27, 2013. https://slate.com/human-interest/2013/12/life-magazine-1945-why-it-was-the-greatest-magazine-ever-published.html. Accessed on 6 August 2024.

Pols, Robert. 2015. *Victorians in Camera: the World of 19th century Studio Photography*. Barnsley: Pen & Sword History.

Pomroy, Matt. 2020. "Why did men stop wearing hats: the decline of male headwear". Esquire Middle East. 20 May 2020. https://www.esquireme.com/style/why-did-men-stop-wearing-hats#:~:text=So%20why%20did%20the%20vast,the%20practise%20began%20to%20decline. Accessed on 6 August 2024.

Povos Indígenas do Brasil. Xavante. https://pib.socioambiental.org/pt/Povo:Xavante. Accessed on 15 September 2020.

Povos Indígenas do Brasil. Xetá. https://www.indios.org.br/en/Povo:Xetá. https://www.indios.org.br/en/Povo:Xetá. Accessed on 26 September 2020.

Pratt, Mary Louise. 1992. *Imperial Eyes: Travel Writing and Transculturation*. London: Routledge.

Prefeitura Municipal de Douradinha - PR. História. http://www.douradina.pr.gov.br/site/prefeitura/cidade/historia/. Accessed on 6 August 2024.

President Nixon Welcomes Leonid Brezhnev to the United States. 1973. https://www.youtube.com/watch?v=B9IcLbgnzfY&t=747s. Accessed on 6 August 2024.

Preto Velho. Wikipedia. https://pt.wikipedia.org/wiki/Preto-Velho. Accessed on 6 August 2024.

Proner, Francisco/Farpa Coletivo. 2018. "Lula é carregado pelo povo após discurso em São Bernardo", 7 de abril de 2018 *Brasil de Fato*, 20 anos. https://www.brasildefato.com.br/2018/04/07/leia-a-integra-do-discurso-historico-de-lula-em-sao-bernardo/. Accessed on 6 August 2024.

Raoni e Megaron na abertura do ATL Rio+20: https://www.youtube.com/watch?v=HSmUvSpBzNE. Accessed on 6 August 2024.

raoni.com. CHIEF RAONI: launch of the "AMAZON EMERGENCY" campaign - episode 3 - Press conference in Paris, Grenier des grands Augustins (atelier Picasso), 30/11/2012. http://raoni.com/news-627.php. Accessed on 26 January 2020.

Raoni Mektuktire. Wikipedia. https://en.wikipedia.org/wiki/Raoni_Metuktire. Accessed on 6 August 2024.

Reijden, Joël van der. 2018. "Le Cercle and the Struggle for the European Continent: CIA, MI6 and Opus Dei Covert Politics", Feb 14, 2018. Institute for the Study of Globalization and Covert Politics. https://isgp-studies.com/le-cercle-pinay#general-vernon-walters. Accessed on 6 August 2024.

Rey de Castro, Carlos, Carlos Laraburre y Correa, Pablo Zumaeta y Júlio César Arana. 2005. *La defensa de los caucheros*. Monumenta Amazónica. Iquitos: CETA.

Reynolds, David, 2007. *Summits. Six Meetings that Shaped the Twentieth Century*. London: Penguin Books.

Riboldi, Bruna. s/d: "Exposição resgata história dos índios no Rio de Janeiro", *Conexão Lusófono*. https://www.conexaolusofona.org/exposicao-resgata-historia-dos-indios-no-rio-de-janeiro/. Accessed on 6 August 2024.

Richard Nixon Foundation. "Leonid Brezhnev's First Visit to the United States", June 18, 1973. https://www.youtube.com/watch?v=idEjhG2NwcI. Accessed on 13 September 2020.

Richard Nixon Presidential Library and Museum. White House Tapes Conversation 943-008, June 18th, 1973. https://www.youtube.com/watch?v=XjKGYBzlp58. Accessed on 6 August 2024.

Richard Nixon Presidential Library and Museum. "Brezhnev's Visit" - Signing Agreements. 1973. June 21, 1973. https://www.youtube.com/watch?v=fBAo2Aky74Y. Accessed on 6 August 2024.

Richard Nixon Presidential Library. White House Tapes Conversation 948-003, June 18, 1973. "Richard Nixon and Leonid Brezhnev discuss their working relationship", June 18, 1973. Accessed on 6 August 2024.

Rogatchevski, Andrei. 2019. "Interpreting for Soviet leaders: The memoirs of semi-visible men", in *Translation and Interpreting Studies.* Amsterdam: John Benjamins Vol. 14:3 (2019), 442-463.

Rohter, Larry. 2019. *Rondon. Uma Biografia.* Translation by Cássio de Arantes Leite. Rio de Janeiro: Objetivo.

Rojas Brown, Ramiro. 2014. "John Brown: un personaje de leyenda y testigo de excepción", in eds. Claudia Steiner Sampedro, Carlos Páramo Bonilla y Roberto Pineda Camacho, in *El Paraíso de diablo: Roger Casement y el informe de Putomayo un siglo después.* Bogotá: Editorial Universidad de los Andes. 253-268

Roland, Ruth A. 1999. *Interpreters as Diplomats: A Diplomatic History of the Role of Interpreters in World Politics.* Ottawa: University of Ottawa Press.

Roquette-Pinto, Edgar. 1935. *Rondônia: anthropologia-ethnographia.* São Paulo: Ed. Nacional. 294.

Russia Today (RT) Question More. 2014. "Interpreter of Khrushchev's 'We will bury you' phrase dies at 81", 16 May, 2014. https://www.rt.com/news/159524-sukhodrev-interpreter-khrushchev-cold-war/. Accessed on 6 August 2024.

Russia Today (RT) Russopedia. s/d. "On this day: Russia in a click: 26 July". https://russiapedia.rt.com/on-this-day/july-26/. Accessed on 6 August 2024.

Russia Today (RT) Spotlight, TV programme. 2009 "General of Interpreters", interview with Viktor Sukhodrev.

Rutler, Fr. George W. 2005. "Cloud of Witnesses: Vernon Walters" *Crisis Magazine*, February 1, 2005. https://www.crisismagazine.com/2005/cloud-of-witnesses-vernon-a-walters. Accessed on 26 September 2020.

Ryan, James R. 1997. *Picturing Empire: Photography and the Visualization of the British Empire.* London: Reaktion.

Safire, William. 1972. "The Moscow Summit", *The New York Times*, July 2, 1972.

Santos, Rita de Cássio Melo. 2022. "Libânio Koluizorocê. Fragmentos da participação indígena na construção nacional", in *Memórias Insurgentes*, v. 1 n. 1 (2022): *Revisitando com os indígenas a formação nacional.* Revista UFRJ.

Santos, Rita de Cássio Melo. "Biografia: Major Libânio Koluizorocê. Brasis e suas Memórias". https://osbrasisesuasmemorias.com.br/biografia-major-libanio-koluizoroce/. Accessed on 30 September 2024.

Scego, Igiaba. (2015/2018). *Adua*. São Paulo: Nós.

Schrad, Mark Lawrence. 2013. "The Vodka Effect: Happy New Year: A short history of booze diplomacy. Politico Magazine. December 30, 2013. https://www.politico.com/magazine/story/2013/12/vodka-russia-foreign-policy-101613_Page3.html. Accessed on 6 August 2024.

Schudel, Matt. 2014. "Viktor Sukhodrev, polished interpreter for Soviet leaders, dies at 81", *The Washington Post*. https://www.washingtonpost.com/world/viktor-sukhodrev-polished-interpreter-for-soviet-leaders-dies-at-81/2014/05/17/674731d8-dde8-11e3-b745-87d39690c5c0_story.html. Accessed on 6 August 2024.

Schudel, Matt. 2017. The story of one of the most famous Russian-English Interpreters. *Dallas Telegraph. The International Newspaper of Dallas*. Jan 27, 2017. https://www.dallastelegraph.com/russian-interpreter-translator-dallas/. Accessed on 6 August 2024.

Segalen, Victor. 1912/1982. *Stèles* 2e éd. critique, commentée et augmentée de plusieurs inédits, Paris: Mercure de France.

Segalen, Victor. 1922/1974. *Le Fils du Ciel*. Paris: GF Flammarion.

Segalen, Victor. 1985. *René Leys*. New York: New York Review Books.

Serva. Leão. 2017. "Sebastião Salgado na Amazônia". *Folha de S. Paulo*. 17 December 2017. https://arte.folha.uol.com.br/ilustrada/2017/sebastiao-salgado/medo/.

Shevchenko, Nikolay. 2020. "U.S. presidents & Soviet leaders trusted this 'man in the middle' the fate of superpower talks", *Russia Beyond*, September 4, 2020. https://www.rbth.com/history/332670-viktor-sukhodrev-cold-war-diplomacy. Accessed on 6 August 2024.

Silva, Carmen Lucia da. 1998. "Sobreviventes do extermínio: uma etnografia das narrativas e lembranças da sociedade Xetá". Dissertação (Mestrado) - Universidade Federal de Santa Catarina, Centro de Filosofia e Ciências Humanas.

Silva, Carmen Lucia da. 2005. "Loureiro Fernandes e os Xetá", em *Arqueologia*, Número especial, Curitiba, v. 3. 197-215.

Silva-Reis, Dennys. 2018. "O intérprete negro na história da tradução oral: da tradição africana ao colonialismo português no Brasil". *Tradução em Revista*, 24, 2018.1.

Silva-Reis, Dennys, and John Milton. 2019. "The history of translation in Brazil through the centuries: In search of a tradition", in *A World Atlas of Translation*, eds. Yves Gambier and Ubaldo Stecconi.. Amsterdam: John Benjamins, 2019. 395-417.

Smith, Carl T. 1985/2005. *Chinese Christians: Elites, Middlemen, and the Church in Hong Kong.* Hong Kong: Hong Kong University Press.

Snapshot (photography). Wikipedia. https://en.wikipedia.org/wiki/Snapshot_(photography). Accessed on 6 August 2024.

Sontag, Susan. 1973. *On Photography.* New York: Farrar, Straus & Giroux. ebook.

Sousa, Fernão de. 1625-1630/1985. "O extenso relatório do governador aos seus filhos", in Beatrix Heintze, *Fontes para a história de Angola do século XVII*, vol. 1. Stuttgart: Franz Steiner Verlag Wiesbaden. 217-362.

South China Morning Post, 1 March, 2011. "Vallon's flight now a distant memory". https://www.scmp.com/article/739494/vallons-flight-now-distant-memory. Accessed on 6 August 2024.

Spence, Jonathan D. 1999. *The Chan's Great Continent: China in Western Minds.* New York: Norton (electronic edition).

Sprague, Stephen F. 2003. "Yoruba Photography: How the Yoruba See Themselves", in Christopher Pinney, and Nicolas Peterson (eds.). 2003. *Photography's Other Histories.* Durham (US) and London: Duke University Press. 240-260.

Stauffer, John, Zoe Trodd, Celeste-Marie Bernier, Henry Louis Gates, and Kenneth B. Morris. 2015. *Picturing Frederick Douglass: an illustrated biography of the nineteenth century's most photographed American.* New York: Liveright.

Steedman, Carolyn. 1986. *Landscape for a Good Woman: a Story of Two Lives.* London: Virago.

Steichen, Edward.1961. "Photography, Men, Sky "To Catch the Instant", *Time Magazine*, April 7, 1961.

Stieglitz, Alfred J. https://www.azquotes.com/author/14143-Alfred_Stieglitz. Accessed on 6 August 2024.

Stieglitz, Alfred J.. 1995. "Alfred Stieglitz: Photographs from the J. Paul Getty Museum", New York: Getty Publications. 82.

Sukhodrev, Viktor. 1999. *Язык мой – друг мой* [Yazyk Moi – Drug Moi], [My Language is My Friend]. Moscow: Olymp, Izdatelstvo ACT. https://www.rulit.me/books/yazyk-moj-drug-moj-read-241350-1.html

Sunburst: https://www.pngegg.com/en/png-fyuwg. Accessed on 6 August 2024.

SYND 18 6 73. 1973. Nixon welcomes Brezhnev at White House. https://www.youtube.com/watch?v=kw5k8tvnRNk. Accessed on 26 September 2020.

Szarkowski, John. 1966/2007. *The Photographer's Eye.* New York: The Museum of Modern Art.

takemethere: My Adventures around the World. Blog by Thomas Carpentier. https://takemethere2.com/2017/08/03/86-train-to-shanghai-city-of-my-ancester-1-may-2017/. Accessed on 6 August 2024.

Taussig, Michael. 2002 *Chamanismo, colonialismo y el hombre salvaje.* Bogotá: Grupo Editorial Norma.

Tavares, Camilo. 1971. *O Dia que Durou 21 Anos*, film by Camilo Tavares, Cidade do México, 1971.

The 61. Wikipedia. https://wikispooks.com/wiki/The_61. Accessed on 6 August 2024.

The Global Medium on the Massachusetts Coast. http://www.stellamaris.no/wnyw3.htm. Accessed on 6 August 2024.

The History Guy. 15 February 2019. The "Goodwill Tour" and 1958 Attack on the Nixon Motorcade. https://www.youtube.com/watch?v=VJhgggC1cmc. Accessed on 6 August 2024.

The Intelligence Profession. Conference Proceedings 3 June 2004. Joint Military Intelligence College. https://ni-u.edu/ni_press/pdf/5482.pdf. Accessed on 26 September 2020.

The Picture Show: Photo Stories from MPR. 2011. 'Smiling Indians' Depicts A Lighter Side Of Native Americans, March 9, 2011. Melissa Block, host. https://www.npr.org/transcripts/134394893?t=1583750802189. Accessed on 6 August 2024.

The Vernon Walters: https://www.google.com/search?q=you+tube-The-Vernon-Walters&oq=you+tube-The-Vernon-Walters&aqs=chrome..69 i57j69i64.2625j0j4&sourceid=chrome&ie=UTF-8. Accessed on 6 August 2024.

The Opium Pipe: Antiques, Collecting, and History. https://theopiumpipe.com/. Accessed on 30 September 2020.

Thomson, John. 1874. Photograph 35, followed by text "From Hankow to the Wu-shan Gorge, Upper Yangtsze". *Illustrations of China and its People: A series of two hundred photographs with letterpress descriptive of the places and people represented*, in four volumes. Vol. 3./ ct3009_1128827. London: Sampson, Low, Marston, Low, and Searle. https://www.loc.gov/item/2021666642/. https://visualizingcultures.mit.edu/john_thomson_china_01/index.html. Accessed on 9 September 2023. Chinese edition: *Illustrations of China and Its People* (2015). Guangxi Normal University Press.

Todze, Tim, dir. 1959. "Khrushchev Does America". Documentary. 59m. https://www.youtube.com/watch?v=VpxsNqAYXB8. Accessed on 26 September 2020.

Torikai, Kumiko (2009). *Voices of the Invisible Presence: Diplomatic Interpreters in Post-World War II Japan.* Amsterdam: John Benjamins.

Trachtenberg, Alan, ed.. 1980. *Classic Essays on Photography,.* New Haven, Conn.: Leet's Island Books.

United States Congressional Serial Set. 1913. *Slavery in Peru.* Washington D.C.

Universal-International News. May 1958 Nixon Ordeal: "Venezuela Mob And Hero's Welcome Home". https://www.youtube.com/watch?v=nvigX1doz2U. Accessed on 6 August 2024.

UOL. "Jornal se desculpa após cortar ativista negra de foto com Greta Thunberg". noticias.uol.com.br/internacional/ultimas-noticias/ 2020/01/ 25/jornal-se-desculpa-apos-cortar-ativista-negra-de-foto-com-greta-thumberg.htm. Accessed on 26 January 2020.

Van der Meer, Hendrika Christina. 1958. *Die Links-Rechts-Polarisation des phänomenalen Raumes.* Gröningen.

Velikaia Rossia. "Nikita Khrushchev in Hollywood". https://www.youtube.com/watch?v=tHNWpgSI_p4&feature=youtu.be. Accessed on 6 August 2024.

V gostiakh u Dmitria Gordona: Viktor Sukhodrev (interview in three parts). 2010. https://www.youtube.com/watch?v=nWa0edoOFpw&t=872s; https://www.youtube.com/watch?v=h7-hw1xd5wk; https://www.youtube.com/watch?v=gTZzHauF3vE. Accessed on 6 August 2024.

Viktor Sukhodrev. Wikipedia. https://en.wikipedia.org/wiki/Viktor_Sukhodrev. Accessed on 27 September 2024.

Viktor Sukhodrev. "Visitando Dmitry Gordon" [Виктор Суходрев. "В гостях у Дмитрия Гордона]. 1/3. 2010. https://www.youtube.com/watch?v=nWa0edoOFpw&t=872s. 2/3. 2010. https://www.youtube.com/watch?v=h7-hw1xd5wk&t=2302s. 3/3. 2010. https://www.youtube.com/watch?v=gTZzHauF3vE

Wallace, Maurice O., and Shawn Michelle Smith, eds. *Pictures and Progress,* Durham, NC: Duke University Press.

Walters, Vernon A.. 1980. *Silent Missions.* Garden City, N.Y.: Doubleday. Portuguese translation by Heitor A. Herreira. *Missões Silenciosas* (1986). Rio de Janeiro: Biblioteca do Exército Editora.

Walters, Vernon A.. 2000. *Poderosos e Humildes.* Translation of *The Mighty and the Meek* by Luiz Paulo Macedo Carvalho. Rio de Janeiro: Biblioteca do Exército Editora.

Wanderley. Andrea C. T. . 2016. Brasiliana Fotográfica. Cartões de visita – cartes de visite, 5 de janeiro de 2016. http://brasilianafotografica.bn.br/?p=3873. Accessed on 20 September 2020.

Weaver, Mary Anne. 1986. "Vernon Walters finds it hard to keep low profile", in *The Christian Science Monitor*, April 18, 1986. https://www.csmonitor.com/1986/0418/overn.html. Accessed on 26 September 2020.

Wexler, Laura. 2012. "'A More Perfect Likeness': Frederick Douglass and the Image of the Nation", in *Pictures and Progress*, ed. Maurice O.,Wallace. and Shawn Michelle Smith, Durham, NC: Duke University Press. 18-40.

Whiffen, Thomas. 1915. *The North-West Amazons: Notes of some months spent among cannibal tribes*. London: Constable.

Wölfflin, Heinrich. 1941. *Gedanken zur Kunstgeschichte*. Basel: Benno Schwabe & Co. Verlag.

Wylie, Lesley. 2013. *Colombia's Forgotten frontier: A Literary Geography of the Putumayo*. Liverpool: University Press.

Zuronskis, Catherine. 2013. *Snapshot Photography: The Lives of Images*. Cambridge, MS: Massachusetts Institute of Technology.

INDEX